the boy who walked

The death of Burton Winters and the politics of Search and Rescue

MICHAEL FRIIS JOHANSEN

Library and Archives Canada Cataloguing in Publication

Johansen, Michael, 1962-, author
    The boy who walked : the death of Burton Winters and the politics of search and rescue / Michael Johansen.

ISBN 978-1-927099-24-7 (bound)

    1. Winters, Burton, -2012. 2. Winters, Burton, -2012--Death and burial.
3. Teenage boys--Newfoundland and Labrador--Biography. 4. Search and rescue operations--Political aspects--Newfoundland and Labrador.
5. Newfoundland and Labrador--Biography. I. Title.

FC2198.3.W45J64 2013   971.8'205092   C2013-906128-2

Published by Boulder Publications
Portugal Cove-St. Philip's, Newfoundland and Labrador
www.boulderpublications.ca

© 2013 Michael Friis Johansen

Editor: Stephanie Porter
Copy editor: Iona Bulgin
Design and layout: John Andrews

Printed in Canada

Excerpts from this publication may be reproduced under licence from Access Copyright, or with the express written permission of Boulder Publications Ltd., or as permitted by law. All rights are otherwise reserved and no part of this publication may be reproduced, stored in a retrieval system, or transmitted in any form or by any means, electronic, mechanical, photocopying, scanning, recording, or otherwise, except as specifically authorized.

 We acknowledge the financial support of the Government of Newfoundland and Labrador through the Department of Tourism, Culture and Recreation.

We acknowledge the financial support for our publishing program by the Government of Canada and the Department of Canadian Heritage through the Canada Book Fund.

table of contents

Prologue:
The unanswered question						7

Part One:
A boy from northern Labrador				15

Part Two:
The anatomy of a search					53

Part Three:
One small boy, one big national debate		81

Epilogue:
A good man gone and other tragedies		185

*I would like to dedicate this book to the
parents and step-parents of Burton Winters,
as well as to the rest of his family and all
his friends—indeed, to anyone who has
lost a loved one and wondered why.*

prologue
The unanswered question

A simple but finely carved wooden cross marks a flower-covered grave in a small cemetery behind the northern Labrador town of Makkovik. No plants have yet taken root in the dry dirt mound and, after a long hard winter, the only blooms that have retained their bright colours are the dozens of plastic ones mourners left behind more than a year earlier. The cross is painted white, like most of the others in the cemetery. It bears no dates, only a name, neatly stencilled in black ink on the crosspiece:

BURTON WINTERS

Had the relevant dates been listed, they would have read: Born July 14, 1998; Died (as far as known) January 30, 2012.

Makkovik's cemetery is in what is known as the Moravian Wood, a small patch of black spruce forest edged by the town, the harbour, and the high bare peaks of Makkovik Hill. The cemetery, the resting place of dozens of the town's citizens, dating back many decades, is bordered by a white picket fence.

The footpath leading to the cemetery begins a short walk away at the community's Moravian Church. More accurately, it starts on the main church grounds behind a modest bungalow known as the Old Manse. Built to serve as the official church residence, it provided housing to Makkovik's permanent Moravian clergy. After the last pastor left, it was rented to a local family—the family of Burton Winters. It was his home for the last few years of his life.

A preliminary incident description from the Joint Rescue Coordination Centre (JRCC) in Halifax relates Burton's final hours in sparse and official diction:

> NFLD EMO requesting assistance in the search for a 14 YOM who ran away from home yesterday. Aurora and Griffon conducted night searches. The snowmobile was located and the surrounding area thoroughly searched but no sign of the search object. Further daytime air search was conducted by Province using CASARA spotters and charter aircraft. Search object subsequently located by RCMP, deceased, approx 8 km from where skidoo was found.

Despite official speculation about trouble between the youth and his parents, such as that alluded to in the incident report, no one really knows why 14-year-old Burton Winters ended up lost on the frozen ocean so far from home. Before his body was located, such speculation may have been necessary as searchers tried to identify a motive for Burton's disappearance. Any insight into what had been going through the teen's mind when he set out into the wintry Labrador afternoon may have held a clue to his whereabouts. Had he gone to visit a friend? Had he just wanted to get away from home for a while?

After the fact, however, talk about hypothetical family arguments no longer serves a public need. Whether or not

Burton "ran away from home"—which few people, if any, believe—remains a private family matter. All that is certain about Burton that winter day is that he turned left when he should have gone right and drove out of Makkovik into a blinding white haze.

Why he did so may never be known, but there are other questions for which Burton's family and their many supporters expect and demand answers: did Canada's search and rescue system fail the 14-year-old, letting him die alone? Could he have been found, and saved, had help come sooner?

Many feel that the officials and politicians, particularly those in Canada's Department of National Defence (DND) and the government of Newfoundland and Labrador, who might have the answers, are evading these questions.

"There are many questions that may never be answered as to why this tragedy occurred, but we have to ensure something like this never happens again," said Jim Lyall of Nain, president of northern Labrador's Nunatsiavut Government (established by the Labrador Inuit at the end of successful land-claim negotiations in 2005) at the time of Burton's death. "We truly believe that Burton would still be with us today if the search and rescue response time had been quicker. We understand the calls for search and rescue were made shortly after the boy went missing, but the air support out of Gander and Goose Bay were not available. That is totally unacceptable."

Burton's paternal grandmother, Anna Jacque, was sitting in her tidy kitchen with her grandson's dog, Quinn, at her heels, when she asked the question that weighed heaviest on everyone's mind: "Why didn't they send the helicopter when they were asked?"

The lack of a satisfactory answer has long rankled with

bereaved members of Burton's family and the residents of his northern community. It resonates with those who live in other isolated communities in Labrador, on the island of Newfoundland, and elsewhere in the Canadian North—anywhere in the country where citizens face weather conditions and geographical dangers similar to those experienced by this lost 14-year-old. The passions of those close to the situation have led to accusations that government officials are evading the truth, hiding a contemptible motive behind withholding aid in the search for an Inuk teen in northern Labrador.

Shortly after Burton's body was found, calls for the federal or provincial government to launch an inquiry into the incident and into the failings of the official search and rescue response began. Why a military helicopter was not provided for the search when one was first requested is at the root of them all.

"Nobody wants to take responsibility for this mistake," Burton's mother, Paulette Winters-Rice, told reporters. "If they [had] told the truth from the start, [the need for an inquiry] would have been avoided. I feel that because he was Inuit, it played a role in what happened and because we are native they figured, 'Oh well, they won't say anything—they're too quiet,' because Inuit people are quiet."

In the days after the tragedy, both Peter Penashue, the Conservative Member of Parliament for Labrador, and Kathy Dunderdale, the Progressive Conservative Premier of Newfoundland and Labrador, said the federal Conservative government should investigate the matter. "My view is that we need to take a look at all of the elements that are involved in search and rescue—right from ground search and rescue, the RCMP, the province, the national defence, all the different elements that are involved in the search and rescue," said Penashue, who was also Newfoundland and Labrador's representative in the federal Cabinet. "Everyone will have to come up with their own conclusions ... We need to improve the process."

About a week after the tragedy, Dunderdale said that her provincial government was seeking better responses from Ottawa: "There are a number of questions that I don't believe we have had satisfactory answers to. The family needs them, the community needs them, and the people of this province need them. I've asked my minister to go to the Department of National Defence and ask for more information."

The premier told reporters that, like most people who had watched the tragedy unfold, she couldn't understand why the military did not immediately send an aircraft to help. "Why not?" she asked. "There may have been very sound reasons for not doing that, but we don't know what they are and we need to know what they are ... People in this province have an expectation and a right to these kind of services and the way they were provided in this case was not satisfactory."

Penashue and Dunderdale softened their demands over time. Most of the politicians who remain committed to a full investigation are in an Opposition party.

"We've got to find out what went wrong," insisted Randy Edmunds, the Liberal member for the Torngat Mountains district in the House of Assembly, and one of the men who found Burton on the ice. "I mean, inquiries have been called for far less [serious] situations than what happened with Burton ... You've got to find out what went wrong. For one reason, if not any other, it's to fix what went wrong. I don't like to see governments blaming each other and the issue not being addressed."

As of late 2013, the many calls for an inquiry have been declined. The Canadian government deemed that it was not necessary, positing that DND had discovered and analyzed all the important information about the matter within days of the tragedy.

"The Department of National Defence will endeavour, as always, to ensure that any lessons that can be learned from this tragic incident are used to strengthen Canada's National Search and Rescue Program," Defence Minister Peter MacKay wrote in a letter to the Newfoundland and Labrador government in early March 2012. "Accordingly," he continued, "an incident report was prepared after a full assessment of all factors, and formed the basis of the February 8, 2012, briefing by Rear-Admiral David Gardam, the Search and Rescue Region commander for the Atlantic region."

Nevertheless, both MacKay and his Cabinet colleague Penashue implied their half-hearted support behind an inquiry—although not a federal one. In May 2012, Penashue made a rare appearance on CBC television to announce that Ottawa would not launch its own investigation, but it would provide information to a provincial one.

"We would not be in a position not to co-operate," Penashue stated. "This is a legally initiated process and everyone would have to co-operate."

By then, however, the provincial government had made its intentions plain. In March 2012 in the House of Assembly, Premier Dunderdale revealed that her government was powerless to call for the type of investigation being demanded: "I have no authority to institute an inquiry into the federal government's activities to have access to the kind of information we would need. I can call on the federal government for such an inquiry—that may very well happen."

Two months after Penashue's offer the province made its position even more clear.

"We do not believe a public inquiry is necessary," Municipal Affairs Minister Kevin O'Brien, responsible for the province's Fire and Emergency Services, wrote to MacKay. "There is ample information in the public domain. We believe a poor judgment call was made, but that does not on its own warrant an inquiry."

So things have stood, from the government's point of view, ever since.

But calls from the public for an inquiry have not ceased. Unfortunately, subsequent emergencies have only added new voices to the chorus that questions the effectiveness of the national search and rescue system—not just in Labrador but across the country.

part one
A boy from northern Labrador

1. The land

On a map Labrador appears as a rough, upright triangle bordered on the south and west by Quebec and on the northeast by the Labrador Sea. A centuries-long boundary dispute between Quebec and Newfoundland (first as a British colony and later as the semi-autonomous Dominion of Newfoundland) formally ended in the late 1920s when the highest court in the British Empire ruled on the issue—although, to this day, the Government of Quebec refuses to recognize the official southern demarcation line.

Except for one 35-year break, Newfoundland governments have held jurisdiction over the territory since the British won it from the French in 1763. The British gave their Newfoundland governor rule over Labrador, even though the Treaty of Paris ending the Seven Years War implicitly considered it to be part of Canada. The agreement doesn't actually name it as a separate territory:

> His Most Christian Majesty [the King of France] renounces all pretensions which he has heretofore formed or might have formed to Nova Scotia or Acadia in all its parts, and guaranties the whole of it, and with all its dependencies, to the King of Great

Britain. Moreover, his Most Christian Majesty cedes and guaranties to his said Britannick Majesty, in full right, Canada, with all its dependencies, as well as the island of Cape Breton, and all the other islands and coasts in the gulph and river of St. Lawrence, and in general, every thing that depends on the said countries, lands, islands, and coasts ...

The Empire's administrators transferred jurisdiction over Labrador to British-governed Quebec just before the start of the American War of Independence in 1776, but the Empire gave Labrador back to Newfoundland 35 years later. Since then, Labrador has been considered an integral part of Newfoundland, although Labradorians were only given the right to vote after World War II, when they were asked to cast ballots for or against Confederation with Canada. Labradorians voted as much as 80 per cent in favour of joining Canada in the two 1948 referenda. Labrador has no special political status to distinguish it from any other part of the province.

Labrador is almost three times the size of the island of Newfoundland, encompassing a little less than 300,000 square kilometres of continental mainland and offshore islands. Sitting on the eastern reaches of the Canadian Shield, this hard, hilly country is replete with innumerable lakes and rivers which flow into either the Atlantic Ocean or a short stretch of the Strait of Belle Isle. Thoroughly scoured by glaciers in the last ice age, most of Labrador's ancient mountains have been worn down into rolling hills, but several lofty ranges remain, including the Red Wine Mountains in the western interior, the Mealy Mountains that shape the long southeastern shores of the Hamilton Inlet, and, highest of all, the Torngat Mountains that form much of the sharp northern backbone of the Labrador Peninsula.

Most of the region, except for the 8,000-kilometre coastline, is a high plateau covered with tundra, bog, and boreal forest. In the milder south, residents enjoy a subarctic climate. Conditions become increasingly harsher the farther north one travels. Beyond Groswater Bay, the climate is almost completely arctic. Hardy species of trees are found in sheltered bays and valleys, but the long cold winters that freeze both the land and the ocean leave no doubt that Labrador is a northern country. It does not easily welcome settlers, and the people who arrived in centuries past, especially those who came to the northern coast, had to prove that they were as tough as the land in order to survive.

2. The people

Just as the current of the Labrador Sea brings frigid water hundreds of kilometres south to freeze the land it touches, so has it brought people from the north able to live in the cold country that they found, making Labrador the home of the world's southernmost Inuit population.

The Inuit first came to the region around 700 years ago at the end of a 300-year-long migration eastward from Alaska across northern Canada. By traditional oral accounts, and confirmed by archaeological research, the ancestors of the modern-day Inuit, historically known as the Thule, were not the first to settle along the coast of Labrador. They were preceded from the north by a people who belonged to the Dorset Culture. Old stories tell of the differences between the two groups and of how the last of the Tunnit, as the Inuit called them, died near Hebron in the fabled past. The Inuit settled in well and by the time explorers, whalers, fishermen, and missionaries came regularly from Europe in the 1500s, they had reached as far south as the northern tip of Newfoundland, where they went to trade, if not ultimately to settle.

The land the Inuit encountered when they followed the southward leading Pebble Islands (known in Inuktitut as Tutjat, or "Stepping Stones") from Baffin Island to the northernmost point of Labrador, today called Cape Chidley, would have appeared as barren and forbidding as any they'd settled across the Arctic. Even more so, possibly, since the stark Torngat Mountains rise straight out of the ocean and offer little or no nutrients or shelter for plants, animals, or humans on their ancient, ice-scoured rocks. However, the Inuit have always been a people more of the sea than of the land, and while the land in Labrador's far north might have appeared bare, the ocean off its shores teemed with aquatic and semi-aquatic animals: fish, seals, walruses, whales, birds, and polar bears in great numbers and variety. Nor was the land truly barren, as it was home to large numbers of rabbit, ptarmigan, caribou, and muskoxen. The farther south the Inuit migrated, travelling by skin boat in summer and dog team in winter, the more abundance they encountered: the granite mountains sunk to become moss-covered then tree-covered hills, the empty valleys between filled with forest and more species of game than they could have possibly known since their forefathers and foremothers left Alaska.

Although the Inuit population in Labrador grew considerably over the centuries, they remained essentially nomadic until at least the 1700s, living in small family groups practical enough to move from one seasonal camp to another with little trouble and staying en route for weeks at a time. Signs of pre-contact Inuit from before the 1500s can be found in every likely place along the shoreline, indicating that they probably spent time in every bay and inlet on the coast, including what is now called Lake Melville.

The most common indications of Inuit presence are overgrown tent rings, but, according to Barry Andersen of Makkovik, travellers might still find the stone *Inukshuk* his Inuit grandfathers built as navigational aids in long lines on the hilltops. In Labrador, these rocky columns were not traditionally built to look like men, as they are in the high Arctic, and were only 1 metre or so tall. Traditional Labrador *Inukshuk* were built to indicate specific routes. All that has changed.

"There used to be small *Inukshuk* with one rock pointing to the trail," Andersen explained. "But everybody is making *Inukshuk* now. Today the old-timers would be going in zigzags."

Prolonged contact with Europeans altered the Inuit way of life, but it didn't happen right away. Vikings from Scandinavia were probably the first Europeans to explore and try to settle northeastern North America, but they came and left long before the Inuit arrived. Chances are the two groups met in Greenland around 1300, just before the Norse colonies died out. The cross-cultural impact of their meeting was likely quite small.

Even after the disappearance of the Dorset, the Inuit weren't alone in Labrador. The Innu, related to the westerly Cree, inhabited and travelled the forest and tundra throughout the interior of the Ungava Peninsula, hunting the same caribou as the Inuit and coming to the coast when necessary. The two existed together for several centuries, sometimes trading and sometimes fighting, and they had no visitors from across the Atlantic until Basque sailors arrived in galleons to hunt whales in the Strait of Belle Isle. The Basque set up summer stations in southern Labrador (Red Bay, the most famous, was declared a UNESCO World Heritage Site in June 2013), where they rendered whale flesh into oil, filling thousands of barrels to take back with them to Spain in the fall.

Even though the Basque travelled to Labrador more frequently, in greater numbers, and stayed longer than the Vikings, they never intended to settle permanently. Like the Vikings, they too stopped returning—but in this case because their intensive hunting techniques had seriously depleted the nearby populations of great whales by the end of the 1500s. This limited the amount of food available to the southern Labrador Inuit. Other than this impact, and the temporary source of iron the Basque gave the local people (the Inuit were accused of burning boats and buildings and scavenging the nails), the Basque probably had little lasting influence on Inuit culture.

By the mid-1700s, Newfoundland fishermen were visiting southern Labrador to establish processing stations (where cod caught offshore was dried and salted on flakes, packed into barrels, and shipped to foreign markets). Some attempts were made to build permanent European settlements—most notably by George Cartwright of England, who, after an initial attempt farther south, was successful in founding a new community at the mouth of Sandwich Bay, a community that still bears his name. Meanwhile, Christian missionaries from the small but innovative Moravian Church were looking to the north. They were more determined than the English trader to remain in Labrador, driven not just by the prospect of commercial gain but by keen religious fervour and a desire to convert the Inuit to Christianity.

3. The church

The first missionaries of the Moravian Church, the *Unitas Fratrum* ("Unity of the Brotherhood"), travelled from Germany to northern Labrador in the middle of the 18th century to build relations with the Inuit. Their initial encounters did not go well, and six of the first missionaries were reportedly killed by a group of Inuit in 1752. Rather than discouraging further missions, this violent incident increased the determination of one Moravian brother to preach among the Inuit and to convert them from their so-called pagan beliefs. As Jens Haven wrote in his memoir:

> In the year 1752, hearing at Herrnhut that Dr. [John Christian] Erhardt, a Missionary sent to the coast of Labrador, had been murdered by the Esquimaux [as the Inuit were known], I felt for the first time a strong impulse to go and preach the Gospel to this very nation and become certain in my own mind that I should go to Labrador.

Haven travelled to Labrador three times, twice to arrange for the British Crown to grant 100,000 acres of the northern colony to the Moravian Church and again to found Nain, the first permanent settlement in northern Labrador, in 1771.

The Unitas Fratrum, one of western Europe's first Protestant denominations, had split off from Roman Catholicism in the late 14th century in what is today the Czech Republic. The Moravians founded congregations that could worship in their own language, with clergy who could marry and who did not preach the severe Catholic concept of Purgatory. After more than 100 years of growth, the Moravians were severely repressed by Catholics for the next 100 years. In the early 1700s, the Moravian Church managed to revive itself in Germany from a small underground remnant of worshippers later called the "Hidden Seed." During that period, the surviving congregations emerged from hiding, established a new community called Herrnhut, and began sending missionaries throughout Europe and to every other inhabited continent. The missionaries sent overseas were charged with establishing communal-type religious settlements that emphasized worship and prayer and ministered to local native populations in their own languages. By the time Nain was founded, the Moravians had sent missionaries to Africa, the Far East, and both Americas. Permanent missions had been established in Australia, in the Danish West Indies, in several locations in what was to become the United States, and in northern Greenland. It was at the Greenland mission that Jens Haven first lived among the Inuit and where he learned to speak Inuktitut.

Nain was not the Moravians' first attempt at founding a permanent mission in the region. Just before his untimely death, John Christian Erhardt had laid the foundations for a settlement in Ford's Bight, a short boat ride from modern-day Makkovik. The colony in Nesbit Harbour, as Erhardt called the bay, was intended to become home for at least 500 families, but the enterprise collapsed following the murder of Erhardt and his colleagues. The one house constructed at the site was destroyed by fire shortly after these men were killed.

After founding Nain, the Moravians established seven more settlements in northern Labrador over the next 133 years, starting with nearby Okak in 1776 and ending with Killinek, in Labrador's far north, in 1904. Five of those settlements were eventually abandoned because of declining populations brought about by disease (the Spanish Flu pandemic of 1918 killed up to one-third of all Labrador Inuit) and by the difficulty of providing public services to remote locations. Most memorably, the residents of Okak (later called Nutak) and Hebron were forcibly resettled and scattered among other Labrador Inuit communities from North West River to Nain by the Newfoundland government in the late 1950s; these relocations were instigated by the local Moravian Church leadership, ostensibly for humanitarian reasons.

Although Killinek was the last mission site to be settled, the youngest of the surviving settlements is the town of Makkovik, which dates its origin to 1859. That's when a Norwegian trader working for the Hudson's Bay Company met a woman from Makkovik Bay and stayed to start a family with her. Many people living in 2013 in this community of less than 400 are descendants of Torsten Kverna Andersen and his wife Mary Ann Thomas; Andersen remains the most common family name in Makkovik. While Makkovik was founded for secular reasons, the Moravian Church soon became an important part of the community, firmly establishing itself in the town by the end of the 19th century with the construction of a large mission complex. As was their practice when constructing buildings for the tree-poor Labrador coast, the Moravians had the complex prefabricated in Germany, taken apart, shipped across the Atlantic Ocean, and reassembled in its new location. Unfortunately, this impressive church building was destroyed by fire after World War II. The Moravians replaced it soon after with another church they transported by boat from nearby Aillek Bay. A third one was eventually built in 1992.

The decades following the 1948 fire brought a gradual decline in the Moravians' importance to Makkovik. As of late

2013, the church still hosted Sunday services, but it no longer had a resident pastor, which is why the house that served as the Manse has been rented out to local families. For several years, the resident family has been that of Rodney and Natalie Jacque—and so the modest bungalow toward the north end of town, with the Moravian Wood behind and a lawn and white picket fence in front, was the primary home of Burton Winters, the place where he should have spent the night of January 29, 2012, instead wandering, lost, on the sea ice.

4. The town

Makkovik is a quiet town. Economically, it is still dependent on the sea, but the fishing business hasn't been lucrative or stable for more than two decades. Fishermen took cod by the tonne from nearby waters until the once-legendary stocks came close to extinction and the fishery was shut down by the Canadian government in the early 1990s. In 2013, about 20 local people fished professionally, mostly for snow crab, but also for turbot. More than 100 locals were employed at peak periods in the two processing plants located between the main wharf and the government's Marine Centre. In 2002, 1.6 million pounds of crab and 500,000 pounds of turbot were processed in Makkovik; since then, the annual catch volumes of both species have steadily declined.

Employment in the community is offered by various government agencies and private businesses. Makkovik has three stores, a hotel with an attached restaurant and several outlets for outside fuel, communications, and transportation companies. The federal government operates a detachment of the Royal Canadian Mounted Police (RCMP), a post office, and a weather station; the provincial government: a school, a diesel-powered electrical generating plant, a medical clinic, and a social services office; the Nunatsiavut Government: a Conservation Office and a Department of Health and Social Development; and the local municipal government has a garage, pump house, museum, craft shop, daycare, and radio station. Not all of these facilities and offices offer much full-time work.

Meeting the needs of tourists and other travellers provides the people of Makkovik with employment and income that could grow substantially in the future. To prepare for an anticipated increase in the number of annual visitors (who must come either by plane, boat, or snowmobile, since no road reaches to the north coast), the municipal government has built long boardwalks to scenic viewpoints around the community and gazebo shelters. The local craft shop sells some articles to tourists, but it mostly provides raw materials to local carvers, artists, and craftspeople, who market their work to outside buyers online. The municipal government organizes an annual July Trout Festival—"Five days of fun, food, and music"—as an attraction for local people and visitors. It also assists outside agencies and researchers in hiring local boats and guides for expeditions into the surrounding wilderness.

On a more traditional economic level, because of their Inuit heritage, the people of Makkovik have the aboriginal right to hunt, fish, and gather food and other resources, such as carving stone, from their lands and waters. Many still net trout and salmon from nearby streams and bays. They hunt migratory ducks and geese and also the sedentary ptarmigan. Seals are hunted on the ice in the spring. For many years hunters went into the interior for caribou, but the decline of the George River herd, once the largest in the world, has made caribou an increasingly rare prey for a hunter on the Ungava Peninsula. Blueberries, partridgeberries, blackberries, and bakeapples, however, are plentiful throughout the wilderness surrounding Makkovik and are picked every autumn for home consumption or for sale.

Makkovik is also quiet socially—at least for those who prefer any kind of nightlife. Although Makkovik is not an officially dry community, it is difficult and expensive for its residents to obtain alcohol. It must be shipped in from the government liquor store in Happy Valley-Goose Bay or a private beer store in Rigolet. Shipments can come on the scheduled coastal boat during the summer, but purchasers face high air cargo rates all winter long. Whether this makes a difference to the quality of life in Makkovik is debatable, but it's clear that the community doesn't suffer the levels of violence and crime experienced in some other northern towns where alcohol can be more readily and cheaply obtained. In fact, when children from other Labrador communities need to be placed into foster care because of difficulties in their home lives, they are often sent to a family in Makkovik. On the whole, Makkovik is considered one of the best places in Labrador to raise a child.

"It is fairly safe for the most part," affirmed Elizabeth Mitchell, principal of the kindergarten-Grade 12 J.C. Erhardt Memorial School and who was born in Makkovik. "There's been an increase in underage drinking and drug use in the community, so it's not as safe as it was. There are more things available to kids, but it's not that bad. It's there, but it's not so bad."

5. The children

Growing up in Makkovik might not be "so bad" for children but it's certainly different than growing up in a less remote community. For example, Makkovik has no cellphone service. People may own mobile phones, but since there's no local tower sending signals over the town, users must go all the way to central Labrador (a 200-kilometre-plus direct flight) before they can call or text anyone. But the Internet is available, and everyone with a computer can get online, although not always by high-speed connection. That, and cable or satellite television, has distinctly impacted the youth of Makkovik.

"It's a whole new world," Mitchell explained. "It's totally different than when I grew up here. We had no TV even … It's different now since computers came. Now everybody has their own TV in their own rooms."

Randy Edmunds, who in 2013 represented the Torngat Mountains district in the provincial House of Assemby, painted a similar picture of growing up in the isolated community: "When we were kids, we grew up in Makkovik and Hopedale—there were no computers and no television. We spent a lot of time outdoors. That's how we grew up, but in the last 40 years it went from how I grew up to the Internet and video games. It gives them [the children of today] a lot more information to absorb and maybe it takes away from their traditional activities."

One thing hasn't changed for Makkovik's youth: they might not get outdoors as much as they used to, but when they do, they go to the same places Makkovik youth have always gone—just maybe a little faster—and they engage in many of the same activities.

"We'd go over to Ranger Bight," Mitchell added, describing an area to the west of Makkovik that encompasses the small bay called Ranger Bight on official maps, the stream that runs into the cove, and all the hills around it. "We'd spend all day at the first pool and the second pool. People still go over there to go swimming."

Being outdoors in Labrador has always been more about survival than play. Before television and computers became a main source of recreation, before snowmobiles arrived to replace dog teams and boat motors to replace sails (and even long afterwards), everyone had to be outside in the short, intense summers to help them make it through the long, cold winters. Berries had to be picked, trees cut, traps set, animals hunted, and fish taken from the sea, streams, and lakes. Today, with easier travel and communication, most people no longer need to go onto the land or ocean to feed their families, but many choose to practise the traditional arts. A general and marked decline in wildlife populations is changing the nature of the woods but, despite an uncertain future, the skills attached to these life-sustaining activities continue to be prized by the Labrador Inuit as both practical and vital parts of their cultural heritage.

In Labrador, as a matter of course, many young people (not just those of Inuit descent, but also Innu, and the later settlers who married into aboriginal communities) learn how to set nets, shoot firearms, and drive snowmobiles. It's not a dry exercise in cultural preservation. They do it because their parents engaged in these activities out of necessity and also because they're still fun.

It's not always easy to keep up with these outdoor traditions and some skills are in danger of being lost unless a concerted effort is made to keep them alive—dog-sledding, for example. No one needs to travel by dog team anymore, but enthusiasts in many Labrador communities continue to practise and teach all the skills needed to raise Labrador huskies and drive teams of them across the winter landscape.

But nowadays most people, young and old, prefer driving snowmobiles, and laws consequently regulate their use. A child must legally be 13 years old to operate one—an age many Canadians might consider to be far too young to drive a fast, heavy machine. But by that time in Labrador, many children are snowmobile veterans. That's because in the Canadian north, a snowmobile is not entirely a recreational vehicle, as it is in most other parts of the country. With so few roads piercing thousands of kilometres of wilderness, and with snow and ice thickly covering the land and water for up to seven months, snowmobiles are an essential means of transportation between communities and out into the wilderness. Although there are cars and trucks in all six communities on Labrador's north coast, none of the municipalities plow their streets in the wintertime. There's no need. These vehicles are parked or stored throughout the snowy months. Snowmobiles and quads—four-wheeled all-terrain vehicles, or ATVs—become the mode of transportation for business or pleasure.

For the younger residents, of course, snowmobiles are mostly about the pleasure. As essential as they are, they're also fun, and youth will gleefully use them as toys—no matter what their horsepower or that they might be risking their own lives. That's why the Junior Canadian Rangers (JCRs) are important: here's an organization that is, in many ways, perfect for the children of the north, teaching them the practical skills of their culture, so that when they're outdoors, they can have fun safely.

6. The Rangers

"Junior Canadian Rangers are proud and skilled youth who are involved in their communities," according to the official JCR pamphlet distributed by DND. "They are girls and boys 12 to 18, who live in remote and isolated areas of Canada that have Canadian Rangers ... The Junior Canadian Rangers Programme is a structured and meaningful enterprise that helps preserve culture and traditions unique to each community ... Junior Canadian Rangers are taught traditional skills, life skills, and Ranger skills. With traditional and life skills included in the curriculum, the community can infuse cultural norms, local language, regional skills, and social needs into the program. It all happens in a fun, friendly and safe environment!"

The JCRs have been popular in Makkovik since the patrol was set up in September 2002. Their program annually attracts several dozen members, which usually amounts to more than 80 per cent of the 12-18 age group at the local school. Some of these youth eventually become full Canadian Rangers, but the JCR program is not designed to recruit youth into the adult patrol in the way that army, air, and sea cadets are often expected to graduate into the full Canadian Armed Forces.

The leader of the Makkovik Patrol of the JCR, Master-Corporal Barry Andersen, is also the Rangers' search and rescue coordinator and counts the town's founding father, Torsten Andersen, as his great-great-great-great-grandfather.

"We fall under the D-cadets, but we're not as rigid as the cadets when it comes to drill," Andersen clarified. "We're more fun than the cadets ... There's no obligation to move on, no recruiting to DND." This fun involves practical training provided by adult members of Makkovik's 19-strong 5 Canadian Ranger Patrol Group, a patrol formed in the community in the late 1940s that often provided guides for people from official agencies and organizations that weren't familiar with northern Labrador.

Andersen's "grandfather used to take Newfoundland Rangers [effectively Labrador's police force at the time] from Nutak to here. It took three weeks."

The Canadian Rangers on the Labrador north coast are the core of the local ground search and rescue teams and are teachers for children who join the JCRs. JCRs are instructed in map reading, orienteering with the use of compass and electronic global positioning systems, first aid, animal tracking, firearm safety, shooting and target practice, outdoor cooking and camping, operating boats and snowmobiles, and even public speaking. It seems to be exactly what the local children want—there's a perpetual list of 11-year-olds waiting to become members when they turn 12.

"All their friends are in [the JRC program] and they speak highly of it so they want to get in, too," Andersen pointed out.

7. The boy

Burton Winters was a JCR, but he joined a little later and perhaps a little less enthusiastically than many of the others in his group.

"He was past age 12 when he joined," Andersen noted. "He enjoyed it after the second year. He became more involved and more active. We had to coax him along. It was just his personality. He was not very outgoing."

Those who knew Burton said he was friendly and imaginative, but mostly quiet, usually keeping apart from the crowd; "reserved" was often used to describe him by both young and old. Other descriptions included "intelligent," "funny," and "loving."

"I picture him every day, from the day he was born up to his death. I wish I could hold him and kiss him," Burton's mother, Paulette Winters-Rice, told reporters. "He was so good."

She described Burton as a teen who enjoyed camping and fishing and who easily accepted other people no matter what their differences. He loved reading books and had been doing so since he was one—likely because she read to him every night before he slept.

"He was a very, very smart student—like an A student," his last school principal, Mitchell, confirmed. "He was a shy student, a bit reserved, but he wasn't arrogant. He had a wonderful sense of humour, a very dry wit … [but] he always was shy. He didn't do a whole lot of talking … He was not popular, but it was not like he didn't fit in with other kids. He had some good friends."

Burton's contemporaries agreed with that description. Because Makkovik is a small community, all the youth knew him to some degree. Dalton Manak, born a year before Burton, is from nearby Postville ("near," in Labrador terms, is only 15 minutes by airplane), but he has lived with extended family in Makkovik for several years. He said that while Burton may not have been the most popular kid in town, that didn't mean he wasn't liked.

"It's not like we outcast him or anything—he just liked to be on his own," Dalton suggested. "He always had a good 'shy factor,' but he was one of those people who make jokes. That's how he'd let people in. He had his own comedy, a dark comedy."

Jacqueline Winters, also a year younger than Burton, and one of his many cousins, said that, despite being so reserved, Burton had a flair for the dramatic that often surprised and delighted her. She and Dalton described the time Burton, who they said normally dressed all in black, showed up at school "with a golden wig and glasses on and a plaid shirt!"

Jacqueline laughed, and continued: "He was shy, but when he opened up he was really funny ... He had his own comedy. If you saw him today, you'd think he was crazy."

Jacqueline and Dalton pointed out that Burton used his ample imagination, in an ironic way, to explore the darker side of his own personality—that's why the black-haired youth liked to wear all black clothing.

"It was a character he made for himself—he always wore black clothing," Dalton remembered. "He presented himself as evil, but through his actions, he couldn't be what he pretended to be." Behind his pretend exterior and even his dark humour, Dalton added, Burton was "a loving character."

That's what Burton's grandmother said as well, which should surprise no one, but her affection for her grandson did not mean her fond description of him is inaccurate.

"He was a loving person," Jacque insisted. "He loved me a lot—me and his grandpa. He loved his dog. He loved Rod and he loved Natalie [his father and stepmother] and he loved Willie, of course."

Willie Flowers is probably the person in Makkovik who knew Burton Winters best. They were not only first cousins (sharing Anna Jacque as their grandmother) but they had also been best friends just about from birth—Burton was born just two months after Willie. They had the same tastes in most things and liked to do just about everything together. "We used to sleep over, play games, go for walks, go for rides, swimming, snowmobiling," Willie listed. "We'd just hang out at his house or at Grandma's … When I was small, we always used to play trucks outside by Grandma's. I used to ride on his machine and we'd go to a place called Tilt Cove."

But for fun, the reserved boy was interested more than anything in computer games, particularly *The Legend of Zelda*.

"He would play it very often and mostly I'd watch," Willie said. "He was good. He beat the game a few times."

Burton's gaming was only one part of a wide range of interests that he avidly pursued. He was developing talents for both music and art and he wasn't shy about displaying them to his lifelong friend.

"[Burton] played piano," Willie noted. "He'd get music from the games and play it on the piano. He played by ear. He liked drawing, too, all freehand. He drew Homer Simpson and Yu-Gi-Oh and *anime*."

Burton's bedroom revealed something of its inhabitant, as bedrooms usually do. It had been kept largely untouched in the weeks following the tragedy, with a sweater folded, books and videos left where he had placed them, his posters on the walls, his television in the corner, his electronic keyboard ready to play, a telescope perched on a shelf, and a Transformers bedspread on his bed. On the outside of the door was a large paper sign lettered, "Burton's room and Quinn's room." On the walls inside were a star chart of the northern hemisphere and map of the world. His books, which rested beside a Simpsons video, included a Stephen King novel, an English dictionary, a novelized adaptation of the board game *Clue*, a guidebook *Creating 3D Comix*, and a slim volume called *Test Your Video Game IQ*.

At the time Burton went missing he was making plans

to follow through with another interest: the performing arts. He wanted to be involved in the upcoming annual Labrador Creative Arts Festival, a drama event for students from across the region that was scheduled to take place in Happy Valley-Goose Bay later in 2012.

"He loved drama," Willie explained. "I'm not sure, but I think that fall there was going to be drama tryouts. He wanted me to try out too, so both of us could go to Goose Bay together."

Burton's main interest was always gaming, but gaming wasn't just a recreational pastime for him. If things had gone differently, it could also have become his professional future.

"I think he wanted to be a game designer," Willie added. "He knew how to do that stuff. He knew the technology."

His grandmother said the same. "He wanted to make his own games," she noted. "That's what he said he was going to do. He had a big interest in science."

Burton's scientific and technological interests were reflected in his schoolwork. Burton started school in Nain, where he lived for a time with his maternal grandmother, but he spent the bulk of his academic years at J.C. Erhardt School in Makkovik. There, he quickly established himself as a good student; less quickly, but just as firmly, he started to show the hidden talents that promised to take him far in life.

"He was just, like, a caring, quiet student who should still be here," Mitchell observed. "We knew of his musical talent and his artistic talent, but we were just starting to become aware of them. He also loved science; it was one of the subjects he excelled at … That year it was almost like it was going to be his year because he was opening up to the staff and students … A lot of the teachers mention how he was coming out more."

Overcoming his natural shyness could not have been easy. The other youth in town may not have shared all his interests, like video-game design, and he did not necessarily share all of theirs, either.

"What's big here is sports, and he was not into sports." Jacqueline emphasized. "He was really different."

Life for most young people in Makkovik, according to Dalton, is still a life outdoors, despite the movies and computers that lure them inside; being outdoors demands something extra: "You could say the norm for boys is to be wild—to go as hard as you can at everything, to be the most outstanding." Burton may not have been fully appreciated because he revealed his talents in more subtle ways.

However, as Mitchell reiterated, shy young Burton was getting to be not so shy. He was growing up, shedding his reserve, and beginning to reveal the man he could have become.

Anna Jacque has three photographs of her grandson sitting on a prominent shelf in her living room—three photos taken at school, one for each of Burton's last three years. She pointed out how much his face had matured in a short time. In the first photo he's a boy. In the third, he's already a young man in appearance. The reason it was happening, she posited, was twofold: he was enjoying an enriching home life with his father and stepmother, who were teaching him many of the skills of his Inuit culture, such as how to bake and cook; and he was benefitting from being a JCR. "He was starting to open up," Jacque added. "He was getting more mature."

Burton's grandmother is not the only person to credit the Junior Rangers for having a positive influence on the teen's life. Patrol leader Andersen said that, while Burton may not have been eager to join the group at first, he eventually warmed up to the experience and was gaining much from it.

"I think his parents signed him up to try to encourage him to become more social," Andersen explained. "He would have rather stayed at home playing video games ... than be out playing with other kids."

It's not unusual for shy individuals to be involved in the JCR and they are never excluded from the activities—quite the opposite.

"We've got three reserved kids in the patrol and we're encouraging them to participate," Andersen noted. "They're there, but they're always off to the side."

After his first year as a JCR, Burton stopped standing on the sidelines and, without coaxing, was beginning to participate wholeheartedly in many of the activities.

"He learned the Canadian Ranger skills: making fires, building lean-tos, identifying different animal tracks," Andersen added. "He was pretty good."

By the fall of 2011, Burton was learning and practising both traditional and Ranger skills. He'd become a good baker at home and was bringing his bread and flummies (a Labrador specialty traditionally baked directly on the top of a hot woodstove) to JCR functions. In the winter of 2011-12, Burton was getting more expert at driving snowmobiles; he had been given a Ski-Doo Tundra 300 by his father, although he still had much to learn about using the machine.

"He was a novice, a first-year driver," Andersen maintained. "According to his father he had never driven before that year. He was okay, I'd say that, but he was not into any challenging conditions."

Burton had already participated once in the JCR's annual two-day field-training exercise for snowmobile operation, but in the first year he had, by necessity, remained someone else's passenger. The weekend-long field-training exercise held on Killman Pond at the end of January 2012 was the first time Burton had attended on his own machine. Since his best friend didn't have a snowmobile, Willie rode along as Burton's passenger.

8. The outing

Willie had joined the JCRs at about the same time as Burton had, and he liked it just as much as his cousin eventually did.

"The Junior Rangers taught us the land pretty good, but only close by," Willie explained. "We have barbecues and outings and in the summer we go to camp. We volunteer in the community and just learn stuff."

The two-day trip to Killman Pond was mostly about snowmobile training, but it also involved setting and checking rabbit snares: "just basic stuff," according to Willie.

The field-training exercise weekend was supposed to have been an overnighter, but because the weather threatened to turn bad, the Rangers went home Saturday evening instead of sleeping in several large tents on the site. They all returned to the pond the following morning. Otherwise, everything went according to plan—in fact, in some ways it went better than they had hoped.

Killman Pond, a small body of water about 6 kilometres south of Makkovik, is a popular destination for people from the community during all seasons. They go there to hunt game and cut firewood or to enjoy an outing on a pleasant day.

"There are trails going in every direction," Andersen said. The JCRs had used Killman Pond once before for a winter field-training exercise, but they hadn't camped in the same place: "We try to go to different locations so the kids don't just get used to one location. It was at the north end of the pond a couple of years ago and we did some fishing. This year [2012] we went across the pond and into the woods 600 or 700 metres."

After their arrival at the pond on January 28, the first day of the field-training exercise, the 18 participating JCRs first learned woodcraft, guided by Andersen and three other adult Rangers.

"On Saturday we drove up to the site," Andersen said. "I was in the lead until we got to the end of the pond. Then they set up rabbit snares and we looked for fox tracks and rabbit tracks to get the JCRs to identify tracks in the snow, for them to know what is game and what is not. We looked for partridge, but we didn't see any."

The Rangers next tried to teach the youth what to do if they experienced one of the most common snowmobile mishaps.

"We taught them about getting stuck in the snow and hauling the snowmobile out again," Andersen recounted. "The best way to get out is to pull by the skis while one guy is gently hitting the throttle. Most people want to push."

When that exercise was done, it was close to suppertime, so some of the JCRs were detailed to cut firewood and others to light a fire. For the sake of practicality, they were also taught how to prime and light Coleman cookstoves, which they used to cook hamburgers. The JCRs had brought along five 10 by 12 feet canvas "prospector" tents and portable tin woodstoves for each tent. The tents were never set up—there was no point because of the weather forecast.

On returning to Killman Pond on Sunday, the JCRs first checked their snares. They caught four rabbits, which they were taught to skin properly. Fortuitously, a local man who had been tracking a moose through the area—the JRCs were aware of his presence and had stayed well away—successfully shot the animal; they helped him field dress the massive carcass.

"They assisted someone paunching his moose," Andersen stated. "A lot of them had been hunting since they were kids, but there were a couple of them like Burton and Willie whose parents didn't hunt."

According to Andersen, lack of hunting experience set the friends apart from their peers, but Burton was already bridging the gap, partly with help and partly through his own efforts. As for Burton's state of mind during that weekend, Andersen hadn't noticed anything amiss: "There didn't seem to be anything with him out of the ordinary. He didn't talk a whole lot, but I didn't notice anything wrong. He was probably more open. He actually sat down and helped cook dinner: hamburgers on a camp stove."

Photographs taken during the event show Burton as he was often described: a tall teenager at the side of a group and almost always with a big smile on his face—or else "making one of those faces he always made," as Andersen observed, flipping through images on his computer screen and spotting Burton's comic grimace.

When the weekend's training exercise was finished, everyone left Killman Pond to converge at the muster point where the JCRs always met before leaving and after returning to town when they went on an outing.

"When we came back from the lake, all the kids had to come for roll call at the new rec centre by the blue sea can [a large metal ocean shipping container]," Andersen continued his narrative. "Everyone answered the roll call. That was at a quarter to 12. Then all the kids went running and jumping onto their snow machines. They all had a half tank of free gas in their machines, so they took it and burned it off."

When youth participate in JCR activities, everything they need is provided, including the standard dark green uniform consisting of track pants, hoodie, and cap with the colourful JCR crest, which displays three green maple leaves on a red background. When they practise shooting, the JCR program provides them with rifles, ammunition, and targets. When they're learning camping skills, they sleep in JCR tents and use JCR stoves and cooking gear. When they are learning safe snowmobile operation, the youth who don't own approved helmets are loaned ones by the program. Every snowmobile needed for the training exercise has its gas tank filled to capacity at the start of the weekend. Much of that gasoline, however, wasn't used during the field-training exercise. The borrowed helmets, tents, and woodstoves were all returned at the end of the exercise after roll call—but, of course, it would have been nearly impossible and quite unsafe to give back the remaining gasoline. The participants were allowed to keep the fuel they hadn't used and, with a whole Sunday afternoon ahead of them, almost every JCR in Makkovik was eager for a joyride around and out of town. So joyride they did, heading out to burn off the windfall of free gasoline. Their tracks went all ways: to Tilt Cove, Ranger Bight, and down below Astronaut Hill.

Everywhere Makkovik youth have always gone when they wanted to have fun with their friends, that's where the JCRs went that day. They all did: everyone together in small groups of friends. Everyone, that is, except Burton Winters. He wound up by himself and, intentionally or inadvertently, he ended up where he'd probably never been before and where his contemporaries usually don't go.

Willie said that when they left the roll call (Burton still driving his Tundra 300, with Willie behind him as passenger), they sped through town to the old Moravian Church Manse where Burton lived before going back to Anna Jacque's house, where they arrived a little after 1 p.m.

"We just got back from the Junior Ranger activity and I was asking if he could give me a ride to Grandma's," Willie remembered. "We stopped to his house [first] because he had to drop off his grandpa's gas can."

Burton's stepmother was home at the time.

"He came home with the rabbit head," Natalie recalled. "I said, 'Burt, I thought it was the rabbit foot that was lucky,' and he said, 'The rabbit head is lucky, too!'"

With Willie waiting outside, Burton couldn't stay long.

"So he said he was going over to his grandma's house for the afternoon," his stepmother recollected. "He usually goes over to grandma's—grandma and grandpa's—and he said, 'Do you want me to put Quinn outside?' And I said, 'yes'—and that was the last I heard of him."

After finishing their errand to the Old Manse, Willie said the two drove back through Makkovik again along the harbour-side Moravian Street, up Harmony Road, turning at the long Big Land grocery store before joining Andersen Street and pulling into their grandmother's yard in front of the west-facing door. Burton only stopped long enough to let him off the snowmobile and he didn't take time to shut off the engine or come inside. Instead, he continued around the house to a small laneway, and from there went back onto the snow-covered street.

By then Willie was indoors. He had the impression Burton intended to drive back to his father's house and he was fairly sure he heard his cousin heading down toward the harbour and Moravian Street. Willie had no idea Burton had any plans to go anywhere but home, and he didn't know why he decided to go out of the town instead: "Maybe he just wanted to go for a ride around the hill. The weather was kind of bad, so maybe he couldn't see the turn. During the Junior Ranger activity it [the weather] was good, but it was starting to get bad. First it started out with light snow and then it got kind of blinding. It was cloudy, but still bright."

9. The ordeal

After dropping Willie off at their grandmother's house, Burton did not turn right to go back down to Moravian Street. He may have instead turned left up Andersen Street, a route that would ultimately have led him over the hump of the town past J.C. Erhardt School and then down toward Ranger Bight—according to their grandmother, Anna Jacque. Her husband, Burton's grandfather, was in a street-facing room when Willie was dropped off, and he had heard the sound of the snowmobile through the window as it passed that side of the house. He said Burton's machine had been moving westward.

If that is indeed the way Burton had gone, he could have made it to Ranger Bight in a few minutes. Once there, he could choose between going left up the stream to the two Ranger Bight ponds—he would no doubt have been able to see that many other snowmobiles had already driven that way—or veering right onto the sea ice in the Bight, then following the shoreline several kilometres around a sharp point of land before heading home into Makkovik Harbour from the north. Burton clearly did not go left; most people think he deliberately went right—not to run away, or to put himself in danger, but simply to burn off his free JCR gasoline by exploring a route home that he had never before taken.

"First he went off towards Ranger Cove and then maybe he planned to swing around the point and come back to his father's place that way," explained Terry Rice, manager with the Makkovik Inuit Community Council. Rice remembered clearly the weather conditions that fateful day and shared a commonly held theory about Burton's route: "It was overcast with a bit of snow, but glaring bright. It was hard to see. It was worse than dark. He went where the Junior Rangers didn't often go. He could have been unfamiliar with the point. Others have missed the same turn—older, more experienced people."

Randy Edmunds painted an almost identical picture.

"He would have come down on the back of this mountain here [Makkovik Hill] following this Ski-Doo track, but in actual fact he was between 50 and 100 metres to the north of the track so, given the conditions, he couldn't see it," Edmunds told CBC television. Edmunds pointed out a well-used snowmobile trail that was skirting the frozen shore of Makkovik Harbour. "This track leads to Makkovik, but it's almost a U-turn at the end of that point [Point of the Bight] where you've got to turn to come into Makkovik and where he wasn't on the main track, the one you see here, and with goggles on and the snow conditions he saw that land over there [indicating the far shore of Makkovik Harbour at the base of Astronaut Hill] and drove right past Makkovik and on the other side is where he would have gotten into the cracks in the ice and the cracks of open water ... and then you're out in the ocean."

The first of the many snowmobile tracks discovered and followed on the land and the sea during the frantic search for Burton may support this theory. Hunters found the track as they were returning home later that Sunday afternoon from a trip north to countryside around Aillik Bay. They came across the track as they traversed the ice- and snow-covered Makkovik Bay toward the Point of the Bight, which leads, and which should have led Burton Winters, into Makkovik Harbour and the town itself.

Andersen, however, was not convinced that track was made by the lost teen: "According to the hunters the tracks were going out to sea. But we didn't know where he went out to sea."

This track first appeared out of the drifting snow cover less than 1 kilometre from Big Island, the easternmost point of which is called Red Cliff and is due north of the mouth of Ranger Bight. The track ran in a northeasterly direction up the middle of Makkovik Bay for a short while before veering increasingly harder to the right, generally toward the northwest skirts of Astronaut Hill. It vanished after a little more than 1 kilometre, well short of Ford's Bight Point. It was either buried under windblown snow or it had been blown away.

If that track was Burton's, it means he had missed seeing the Point of the Bight, the landmark that should have guided him back to town, but he probably missed it by much more than 50 or 100 metres. If it was his track, he was almost 1 kilometre from land. It would mean that if Burton had been on Ranger Bight and wanted to follow the shoreline around the point to the harbour, he must have either purposefully headed to Big Island, which is possible, but it was 2 kilometres away and would likely not have been visible to him from the Bight, or he had lost the shore almost immediately as he headed north and the island's Red Cliff point might have been the first land he happened upon in the snowy glare.

If Burton had been out that way before—this is not likely, since his previous snowmobile trips were usually to the southeast onto the land to cut firewood with his father—or if he'd learned the area from a map, as he might have done while in the JCRs, he might have misjudged the distance he'd already driven over the ice, as he might have been driving much faster than he'd realized, and imagined himself still to be near Ranger Bight Point, not 1 kilometre out in the open bay. If Burton had made the track the hunters discovered, he may have still been thinking he knew where he was going. He may have been hoping he was veering toward the Point of the Bight, but instead he eventually found Ford's Bight Point. He'd found the way to the open sea, not the gateway to Makkovik Harbour.

"To speculate: it was snowing, it was dull, but it was not bad weather—I'm speculating that he went up on land and wanted to go around on the ice back to Makkovik, but he couldn't." Edmunds sympathized with Burton's plight. "I've been lost, when you don't know which way to go."

Two more tracks that might indicate where Burton had driven his snowmobile were found farther out toward the ocean, both in the vicinity of Strawberry Head, which is across the bay opposite Ford's Bight Point. Those tracks were already in an area of broken and shifting pans of ice. One of the tracks, the only one that Andersen was certain had been made by Burton's snowmobile, was extremely short and virtually invisible, lasting less than 2 metres and almost completely obliterated by drifting snow. The other track, which Andersen was not convinced was Burton's, led to a wide section of dangerous open water, a crack of water so full of slush and shards of ice that it could easily have looked solid to someone who was not paying attention, or who could not see ahead with any clarity. This track appeared to suggest that someone had plunged, machine and all, into the dark, cold depths.

Andersen stressed that any confirmation of where Burton had gone only came with hindsight. As searchers found different tracks, they had no way of knowing who made them. "The snowmobile was the first definite sign"—meaning that until they found the machine, there had been no way to know the boy's direction of travel, or even whether he was on land or sea.

Burton's Tundra 300 was discovered a short distance along the shore of Cape Strawberry off Skipper Cove Point. Burton had driven the snowmobile up into a craggy wall of broken ice and gotten it jammed between the massive slabs. It's quite likely, Andersen said, that because of the weather conditions, the teen simply could not have seen the obstruction in front of him until it was too late and he had run up and into it; the route he had picked ahead of his skis may have looked like flat ice. Sea ice, especially within sight of the open ocean and within reach of the ocean swells, is always in motion. The ice that forms as thick shorefast sheets breaks apart to create open leads, or it crashes together to thrust sharp ridges skyward, like miniature continents building instant mountain ranges.

Many in Makkovik who are familiar with the winter conditions on the channels around their town are surprised Burton got as far as he had on the snowmobile; the ice ridge that finally stopped him would have been one of many that he would have had to avoid, and he would have had just as many open leads to go around, as well, if not more.

Apparently, all efforts Burton had made to free his snowmobile from the ice wall failed, and he was forced to continue his journey on foot, which is not surprising since it later took three men to pry the machine loose. In the snowmobile's gas tank and in a spare, half-full can of gasoline, he had left enough fuel to drive the machine back to Makkovik, as was done a few days later. Almost every step Burton had taken from that moment on carried him farther and farther from his home.

"Once you're into reduced visibility, what you're seeing is not what you see on a clear day," Edmunds pointed out. "He followed the land and saw some islands and maybe he saw the lighthouse on the islands. For the first 2 or 3 kilometres he was on ice that was moving and he must have had difficulties we can only imagine—lost in those conditions, in panic mode, in the dark. I'm going on my own experience of when I got lost and was choking back panic."

Burton's footprints first led away from the open sea and back toward Cape Strawberry, less than 1 kilometre away.

"He went back ashore to Nipper Cove Point and then he followed the mainland until he got to Foxy Rocks," Andersen averred. "That's when he came to open water."

Burton's footprints indicated that, throughout most of his trek across the shifting ice, visibility was improving; perhaps the blinding haze he'd been in at the start of his journey had dissipated as night set in.

"The tracks didn't look like someone lost to me," Andersen added. "He was following from point to point."

Aerial photographs taken by search aircraft several days after Burton's body was found support Andersen's theory. They show footprints that trace remarkably straight lines across the mouth of the wide bay. Burton must have been able to see the point of land ahead of him at that time and at first he headed for his chosen landmarks with strength and energy. That did not last. Leaving Pomiadluk Point—by then he was almost 15 kilometres east of Makkovik—he stopped skirting the mainland and took a bearing that pointed him out toward the long mass of Kidlialuit Island and brought him back within reach of the open sea. Burton had made it almost 6 kilometres across the broad expanse of the channel before he could walk no more.

"He was walking continuously," confirmed Edmunds, who saw the tracks. "At the end it didn't look like footprints, it looked like skiing. He was dragging his feet. He didn't have any energy left."

Cold kills in stages. Hypothermia progresses through a series of symptoms that usually begin with mild shivering, increased heart rate, and rapid breathing. As the cold penetrates, the body limits blood flow to the extremities, saving most of the blood for the torso in order to keep the vital organs warm. If the body is allowed to cool even more, the shivering becomes more violent and the sufferer loses muscle control. By then, so little blood is allowed to go up the neck or down the limbs that all toes, fingers, ears, and lips become pale and turn blue with frostbite. In the later stages, hypothermia cripples a person's ability to think, speak, or remember, eventually leading to a general state of stupor. Hands become useless and walking beyond a stumbling shuffle becomes more difficult. By that time, in most cases, breathing and heart rate have slowed considerably. In the final extreme of hypothermia, the major organs fail and the body dies. Because of the extreme cold, brain death occurs last, which on rare occasions means that rescuers can save a hypothermia victim who appears on first sight to be beyond hope.

"They're not dead until they're warm and dead," as Andersen said, repeating a standard medical protocol for not immediately giving up. Some people, generally children, who tend to be more resistant to the harsher consequences of hypothermia than adults, have been revived more than an hour after the usually fatal unconsciousness has set in.

Survivors of severe hypothermia often report that, in their last moments awake, they begin to feel warm again, and as death approaches it comes like falling asleep. In many cases, victims are found to have removed items of clothing, as if they no longer feel the cold, and to have pushed themselves into any available small space—a behaviour known as the "hide-and-die syndrome" or "terminal burrowing."

Burton Winters, after walking a total of about 19 kilometres since abandoning his snowmobile in the wall of broken ice off Cape Strawberry, following points of land ever eastward, could go no farther. He had been too cold for too long and his body could no longer keep any part of itself warm enough for life. At the start of his trip he had been wearing outdoor gear, goggles, snowpants, coat, and winter boots. However, at a nameless spot of ice at least 1 kilometre from the nearest land—he was due west of a small archipelago called the Ironbound Islands—he had removed some clothing. He had discarded his hat, his gloves, and his goggles. Then, he had laid down on his back in the snow, and never moved again.

At best estimate, Burton had left Makkovik shortly before 2 p.m. on Sunday and it could have taken him a mere 10 to 15 minutes, if the ice conditions had given him a smooth driving surface, to race his snowmobile to the place where he had gotten it stuck. His trip had lasted several more hours at least.

"A normally healthy human being can walk a kilometre in 20 minutes," Andersen pointed out, thinking about how long it might have taken for Burton to cover 19 kilometres on foot. "So three hours, or maybe double that to six at the maximum." Of course, no one will ever know how long he may have remained with his snowmobile before abandoning it, or how fast he had walked.

The coroner's report from the provincial Office of the Chief Medical Examiner is not available to the public, but by all accounts the actual hour of Burton's death was almost impossible for anyone to establish, given the nature of hypothermia and the length of time it had taken to recover the body.

"I have no idea of the time of death," Andersen said. "Even the coroner wouldn't speculate about it."

Fourteen-year-old Burton Winters, late of Makkovik, had probably ended his long walk over the ice on the afternoon of January 30, 2012, the day after he had gone missing from home, but no one can know for sure when he had finally fallen asleep and when his life had quietly slipped away.

part two
The anatomy of a search

Day one: Sunday, January 29

When Burton Winters didn't show up at the Old Manse in time for Sunday night dinner, his family grew worried. His stepmother took a call from Anna Jacque, Burton's grandmother.

"She usually calls and asks how Eliot is—our baby—and how Burt is doing," Natalie recounted. "And so when she asked how Burt was, I said, 'I thought he was with you,' and she said, 'No, he hasn't been here.'"

Naturally, one of the first people Burton's father and stepmother went to for information was Willie Flowers.

"His father came here looking for him," Willie remembered. "He was wondering if I'd seen him."

That was a little after 6 p.m. Burton's father, Rodney Jacque, was unable to find the teen at Willie's place, with Burton's grandmother, or at any of the other usual places that he frequently visited. He called the local detachment of the RCMP around 7 p.m. to report his son as missing and spoke with RCMP Corporal Kimball Vardy.

"Makkovik RCMP receive a report of an overdue youth travelling on snowmobile," reads the provincially approved timeline compiled from records of the RCMP, the Government of Newfoundland and Labrador, and DND. "He was last seen at 2 p.m. (NL time). A search was conducted in the area during the evening period by the RCMP and local search and rescue members."

Vardy took charge of the situation as search coordinator and immediately informed Barry Andersen, who, among all his other positions, is also a community constable. Andersen had gone home for the evening after a long day in the woods with the JCRs, but he lived only minutes from the detachment building. Vardy and another officer went into the community to track down Burton's friends and acquaintances and interview them about who had last seen him and where he might have gone that afternoon.

"Sunday evening the RCMP detachment members here … came to my house looking for my daughter. They were wondering if [Amanda Dyson] had seen [Burton] since their Ranger outing earlier that day," Makkovik resident Perry Dyson told CBC. That was the first Dyson had heard about Burton's being missing. He soon became one of the dozens of citizens involved in looking for the teen—about 60 people from both Makkovik and Postville took part. Like many, Dyson had a personal stake because Burton was a friend of his daughter's.

As Dyson explained, "Amanda grew up with Burton. They were in the same class, born weeks apart. They [the RCMP] were looking for Amanda, of course, as I mentioned. She hadn't seen him and she was almost immediately, you know, you could tell she was a bit concerned about that. I told the boys that I'd be glad to help if there was some way I could do so. About a half hour or so later they called to ask if I could take a ride around to some of the cabins in the closer areas to see if he had maybe happened upon a cabin somewhere. So myself and my brother Travis took a run over to the north side of the bay ... but we found nothing, of course."

Willie, who had left his house that evening after supper, encountered the police in those first minutes of the search. He, too, was getting worried. "I was riding around with some friends," he stated. "The cops stopped us and they asked if I'd seen Burton that day. I said, no, not since dinnertime." Willie returned home to wait and to hope that he would soon hear good news. At the time he assumed that Burton's Tundra had broken down or run out of gas.

Once the RCMP's house-to-house search began, word that a local youth had gone missing spread quickly. By then the official search was well under way, but an unofficial one had also started. No private citizens were supposed to look for Burton without direction from the official search coordinators, but that didn't stop some of the community's youth from trying to help out on their own.

"The kids weren't allowed to look for him, but we loved him so much we had to," explained Jacqueline Winters.

"Everybody went back looking for tracks," added Dalton Manak. Several JCRs took to their snowmobiles again to look for Burton's trail up Ranger Bight Brook and out around Tilt Cove, where they had been burning off the free JRC gasoline earlier that afternoon. They did not find any sign of Burton or his Tundra 300.

Meanwhile, Andersen and Vardy organized a nighttime search of the town and its immediate vicinity; they also prepared for a wider operation. By about 10:40 p.m. they requested an aircraft to come to Makkovik to help.

"RCMP Makkovik contacted RCMP Operational Support Services in St. John's requesting air support," the official timeline reports—a timeline, it is worth noting, written specifically to document this particular action. "Through discussions with

RCMP Makkovik, it was decided to have the search continue in the community. Based on the investigational findings, the ground search teams continued to follow leads and ensure a thorough search of the community and immediate surrounding area was completed considering all investigation information available. Searchers did not have a starting point and continued to look throughout the community and surrounding area. Present weather conditions were deteriorating."

The clause "searchers did not have a starting point" meant that no one had any idea where Burton might have gone. According to Andersen and others, until almost midnight they had no clues about any possible direction the youth may have taken—assuming that he'd even left the community. As part of the initial ground search around Makkovik, they probed every available avenue.

"We started canvassing door to door," Dyson explained. "We went off in tangents looking in peoples' woodpaths and stuff in hopes that maybe he got into somebody's woodpath and gotten stuck or something like that, but we had no success that night."

MHA Randy Edmunds also became involved in the search that first night, although he wasn't in the community when it had gotten under way. He happened to be driving his own snowmobile home from the neighbouring community.

"I was in Postville until Sunday afternoon," recalled Edmunds, who owns the only hotel and restaurant in Makkovik. Burton knew the establishment well, as he hung out at the restaurant with the other local youth. "When I got back to Makkovik, it was snowing. It was dark when I left, and I got to Makkovik about 9 p.m. Someone asked, 'Did you see anyone?' Then I got the keys to the hotel. You've got to check as many buildings as you can."

The first clue to the route Burton had taken after leaving his grandmother's house was a snowmobile track that led east up the middle of Makkovik Bay away from Big Island, indicating that he might have gone out of Makkovik, as feared. The hunters who had spotted the track earlier in the day had not thought it important enough to report to the search and rescue coordinator until 11:45 p.m., after they learned that Burton was missing.

"A community member attended the detachment and advised that he had seen a snowmobile track on the ice heading from Makkovik Bay out toward the 'Shima,' the edge of the ice," the timeline states. "The community member stated that he observed the track at approximately 2:30 p.m. (NL time). Searchers were sent out to follow the track to see if they could get a direction of travel."

According to Dyson, the hunters had been curious about the track but did not think too much of it. "There was a couple of local guys who went out for a hunt that same afternoon and they had intersected a track, which really wasn't a commonly travelled track—you know, it wasn't in the area that somebody would normally have been going, had they not been out seal hunting or something like that. That track, really, while as interesting as it was that day, because people may have been out hunting, it didn't really show of any importance to them until later that evening when they realized, of course, that Burton was missing." The track quickly got everyone's attention. Dyson noted that "[a]t that point it was the only possibility we had—you know, we might find which direction he went, eh? Up to that point we had no clue because it was so uncharacteristic of him to venture off like that."

Andersen was among the searchers who went out on the sea ice in the dark to look at this first potential clue. They found the track, as reported by the hunters, and traced it eastward past Makkovik Harbour. They tried to keep driving in the direction of the track, generally toward Ford's Bight Point, but poor sea conditions forced them to stop before they got very far beyond where the snowmobile trail disappeared under blowing snow.

"That's where we turned back Sunday night," Andersen said, indicating the channel off Ford's Bight Point, which has a clear line of sight to the open ocean and is subject to ocean swells. "The ice was going up and down with the waves."

Dyson described the steadily worsening ice and weather conditions that severely stymied their early search efforts on Sunday night: "It was starting to snow that evening, so the track was getting a little harder to follow. That time of the year, even the ice was moving a bit. In the dark and snow it wasn't safe to continue out any further than they did that night."

That first party returned from Makkovik Bay just after midnight, marking the beginning of the second day of the search.

"The searchers turned back after getting as far as possible but had to return due to poor ice conditions," the official timeline reports. "The searchers could not confirm track origin or direction of travel due to weather conditions."

The searchers were uncertain where their first clue might lead them, but at least they had something to follow.

"We really were searching all over: east, southeast, north, and west," Andersen described, "until we came up with the snowmobile tracks leading towards the ocean."

Day two: Monday, January 30

The first full day of the search did not get off to a good start.

After the first search team was "debriefed on track location," a few more people were given radios, ropes, and floater suits and sent back out onto frozen Makkovik Bay. Unfortunately, the second party couldn't get much farther than the first.

"Search team reports track heading toward open water," the timeline reads. "Bad ice is reported making it too dangerous for search, especially at night."

By 2:20 a.m. everyone had to give up and return to land, but only for a few hours. "All remaining searchers returned and the operations were called off for the night. Searchers arrange[d] to meet at 7:00 a.m. to continue search," reads the timeline.

A call from police officials in St. John's came a little after 6 a.m.

"RCMP Operational Support Services contacted RCMP in Makkovik for an update," the timeline continues. "Burton Winters had not been located. Makkovik RCMP advised that the snow continued to fall and covered any tracks."

The volunteers who had been involved in the search the previous night, as well as some additional ones, met as arranged at the RCMP detachment at 7 a.m. and went back out through the town to cover the ground they'd already examined the evening before, in case darkness had prevented them from seeing something, anything, that might have been important. They also expanded the search area—not just onto the sea ice, but farther out on land as well, since they as yet had no proof that the track the hunters had found the day before had anything to do with Burton Winters.

"All the searchers you saw out there on the ice was only one-third of the searchers," Andersen said, describing the scope of the operation. "All the others were inland."

Edmunds detailed the situation a few days afterwards to CBC: "Really, they didn't have anything to go on Sunday night.

They did find a snowmobile track and the search started, but that night and pretty much all the next day the search effort was kind of hampered by the weather and, you know, we did manage to get a helicopter from Woodward's Oil and they took part initially and another chopper came up. In the meantime we had search parties out from the communities of Makkovik and even the ground search team from Postville assisted, so there was no shortage of local support and searchers."

Efforts to get an aircraft into the area to aid in the search resumed as the ground teams returned to the land and onto the sea ice. Vardy's request to RCMP Support Services the night before had resulted in a St. John's officer calling the provincial emergency measures organization, Fire and Emergency Services Newfoundland and Labrador (FES-NL), at 7:49 a.m. According to the timeline, the following actions occurred between 8 and 9:08 a.m.:

1. FES-NL returned the call to RCMP. RCMP outlines situation with the ongoing search in Makkovik. RCMP relays formal request for air support.

2. FES-NL contacts Government Air Services (GAS) in Gander—outlines the situation and requests availability of helicopter. GAS advises they will need to contact provider (Universal) and will advise shortly.

3. GAS contacts Universal (Goose Bay) and requests dispatch of helicopter.

4. Universal (Goose Bay) checks Environment Canada weather and places call to Postville to determine local weather conditions on the coast. Universal dispatch consults with pilot.

5. Universal calls back to GAS to advise that weather conditions will not permit dispatch from Goose Bay to Makkovik.

6. GAS contacts FES-NL to advise helicopter could not fly due to weather conditions.

7. FES-NL contacts RCMP and advises that contract helicopter cannot fly due to weather. RCMP confirms request for air support and it is agreed that FES-NL will contact JRCC with a humanitarian assistance request for air support.

8. JRCC received first call from FES-NL to request assistance in locating a missing person.

The Joint Rescue Command Centre (JRCC) in Halifax, according to its incident log, opened a file on the Burton Winters case at 1309 Zulu (9:09 a.m. Labrador time) on Monday, January 30, and, four minutes later, recorded the call from Paul Peddle, SAR coordinator with Newfoundland and Labrador's emergency measures organization: "Yes, good morning. It's Paul Peddle calling from Fire and Emergency Services over in St. John's ... the weather is down in the area and we can't get a small chopper or plane in the sky to head to Makkovik," Peddle told the JRCC officer. "I don't know whether ye can do it or not, but wondering if you can do a humanitarian mission, go have a look and see if you can find the young fella."

As recorded in the log: "Requesting our assistance in the search for a 14 YOM [year-old male] who left on his skidoo after an argument with his parents ... They have searched the community and cannot locate him ... weather in area will not permit launch of local helo or aircraft ..."

Within the next few minutes, the JRCC controller on duty, who is only identified as C. Macdonald, had learned from the officer in charge that none of the Cormorant aircraft stationed in Newfoundland would be made available to help. The officer in charge "would only be willing to commit 444 Sqn." The 444 Squadron is a Canadian Air Force unit that flies out of the Goose Bay air base in central Labrador. The squadron's two Griffon helicopters have secondary search and rescue responsibilities, but they are primarily tasked for operational support.

The Bell CH-146 Griffon is a multi-function utility helicopter designed for reconnaissance, for providing aerial firepower, and for search and rescue. The 17.1-metre-long Griffon has a flying range of 656 kilometres. It holds a crew of three, who are equipped with Generation III Image Intensification Night Vision goggles, and it has a Wescam 16D-A Thermal Imaging System.

Macdonald contacted Goose Bay Ops by phone at 9:19 a.m. None of the commanding officers had arrived at the office for the day, but the JRCC dispatcher did receive some good news, if short-lived.

"They are reporting SAR ready," he wrote.

From Goose Bay Ops, Macdonald also obtained telephone numbers for two main 444 Squadron officers. He reached Captain Dan Gillis within 10 minutes and received the real news—the bad news. "[Gillis] is on the way to work but the AC [aircraft] is US [unserviceable] and he will call me when he gets to the Sqn," Macdonald wrote.

Two minutes later Macdonald was on the phone again with his officer in charge (OIC): "Discussed weather 600/1 and aircraft status for the fleet with the OIC. At this point he does not want to commit resources other than 444 and they are US."

Immediately after that Macdonald informed FES-NL that, because of the bad weather and inconvenient "aircraft status," DND could not yet support any search in northern Labrador. He gave them some assurance of possible forthcoming aid: "We may be able to support in the future if required and aircraft/[weather] situations improve."

More than an hour later, the 444 Squadron informed Macdonald that one of the two helicopters being serviced in Goose Bay would likely remain out of commission until 2 p.m. The other would not be available at all—it was in pieces on the hanger floor in the middle of an extensive maintenance operation. That call with the 444 Squadron was the last action recorded in the JRCC log for January 30 until Macdonald added a final note declaring the Burton Winters incident "CLOSED" at 2118 Zulu (5:18 p.m. Labrador time).

Consequently, the first aircraft to join the search for Burton was a private, uncontracted helicopter owned by Woodward's Aviation of Happy Valley-Goose Bay. The pilot, who had spent the night in nearby Postville, decided that the bad weather had lifted enough to allow her to fly out at 10 a.m., although the trip to Makkovik took a little longer than normal. She landed there 40 minutes after taking off and, as arranged, picked up three searchers and began a flight over Makkovik Bay.

On learning from the RCMP that a private aircraft was joining the search, FES-NL officials tried to find out if the Universal helicopter they wanted to charter out of Happy Valley-Goose Bay could also get off the ground and fly to the coast.

"FES-NL officials receive a call from RCMP in St. John's who had been in contact with the RCMP in Makkovik who advised them a private aircraft was about to land in the community," the timeline reports. "RCMP in St. John's asked if it was now possible to fly in air support. FES-NL contacted Air Services. Air Services spoke to the contracted pilot who indicated he would attempt to fly to Makkovik even though there were still weather concerns."

The Universal helicopter left Goose Bay just before 11 a.m., arrived in Makkovik about an hour later, and began the search shortly thereafter.

That morning, the weather was of great concern to all the searchers, both in the air and on the ground. Andersen was one of the three local men—along with Perry Dyson and one other—on board the Woodward's helicopter to scout along the ice farther out Makkovik Bay than they had been able to reach when travelling by snowmobile the night before.

"It was snowing quite thick out to the east," Andersen said afterwards. "I was in the helicopter during the first initial search with the helicopter from Melvin Woodward's oil company. When we were flying out Strawberry Head, heading east, it was just like a wall. It looked like ice fog out towards the sea. I could see no landmarks out there—no islands or anything."

This search party soon ran into trouble, but not because of the weather.

"Private helicopter is forced to land due to mechanical problems," the timeline states. What followed illustrates the constant danger faced by those who engage in wilderness search and rescue operations: the searchers can easily become lost as well. In this case, Dyson and Anderson barely escaped plunging through thin ice into the frigid Atlantic

Ocean. Andersen said it was difficult to tell the difference between solid ice and treacherous snow. "We were just confirming some tracks when I went down. My one leg went down into the water and the guy behind me went up to his armpits. The ice was all broken up and we just stepped on the wrong spot. Fifteen centimetres of snow fell that night and it was all covered over, all the same level ... we were walking along fine and the next thing you know you're up to your armpits. Perry pulled me out."

Dyson happened to be behind Andersen. He remembers it a little differently, giving some of the credit for *his* rescue to Andersen.

"Myself and Barry and another fellow ... went by chopper," Dyson explained. "Some other guys followed the ... the snow-machine track out on the ice on Ski-Doo. We all went out the bay to Cape Makkovik ... It's an area where the sea heaves in as rough as it does anywhere. So the ice is pretty treacherous. It gets soft around the shore from all the rising and falling with the sea and the tides, especially at that time of year when there was so little snow around the shoreline. Some of the guys that were on foot, they went out there on snow machines and they put their snow machines up against the shore in safety and they continued out on the sea ice on foot towards the area of open water. The senior man there was Perry Voisey. He had found what he was confident was snowmobile tracks. He wanted someone to come and see it and, as I mentioned before, Barry, myself, and Errol went out on helicopter and she had developed mechanical issues. So we had actually landed the aircraft very hastily on top of Strawberry Head. We could see Perry down on the ice and he was motioning for someone to come and see ... so Barry and I went down.

"We were walking out to where Perry was hovering around the snowmobile track he was so adamant was indeed there and Barry was walking forward and I was following behind Barry and he stepped in one spot and he actually broke through. I said, 'Oh Jeez Barry, you broke through!' I stepped around it and I broke through and I fell in. Barry very quickly he turned around and grabbed me and hauled me out of there."

The saltwater dunkings did not deter either searcher from continuing, even though crippling frostbite was always an imminent danger. Dyson dismissed the incident afterwards, explaining that he had been dressed properly for the weather and that the air hadn't been cold enough to make him worry about wet socks. "That stuff happens," he said.

The snowmobile track Voisey had discovered took on great importance for the searchers.

"We followed this track that day and it lead to a hole," Dyson clarified. That hole became a distraction; for a while it seemed the search operation would end at its broken edge. "It was a huge crack, a long crack, which was probably open I'd say 50 or 60 metres, maybe. So then the next step was, of course—if the snowmobile track leads to open water then the snowmobile track has got to be in the water, the snowmobile has to be there. Perry and his boys hauled out a boat and he started dragging the area."

According to the timeline, the hole was found at 12:20 p.m. and a boat was put into it shortly after 3 p.m. The men in the boat were equipped with grappling hooks and anchors and used them to drag the sea bottom in an attempt to snag the snowmobile or the person they suspected was down there.

"Search teams discover snowmobile tracks heading for open water," the timeline continues. "A request for an underwater camera is made. Several search team members return to Makkovik to retrieve a boat for a water search. Remaining members stay on scene, sweeping the area. Contracted helicopter continues to search the area."

In a later entry, the RCMP provide an update: "Search team is on site with boat. The team cannot locate corresponding track on the other side of open water or any other debris. Conditions worsen, with heavy flurries. Contracted helicopter is required to leave the scene prior to nightfall to return to Goose Bay."

All the searchers around the gaping hole, as well as those elsewhere on the ice and land, were forced to give up their search that day almost two hours after the second helicopter left.

"Search parties return to Makkovik," the RCMP report in the timeline at 4:39 p.m. "Conditions are too dangerous for night search. Plans are made to commence at first light."

Day three: Tuesday, January 31

The second morning of the search dawned colder than the first, and it was still snowing. Search coordinators sent ground parties out on snowmobile to Monkey, Big Bight, Sharp Hill, and Adlavik Bay, but they all returned empty-handed by noon. None had discovered anything of importance.

"Tuesday morning four of us went south checking cabins," Edmunds recalled. "One group was concentrating on the track that looked like it went into open water and a request went out for underwater equipment."

While the searchers waited for the equipment, which wasn't flown in until later in the afternoon, they prepared for its arrival. All the while, they fought off the dread that seemed to emanate from that dark, watery hole. Even the senior RCMP officer on the scene let himself be overtaken by pessimism as he spoke with the media.

"Well, right now we have ... searchers are probably going to be going out today and checking around the water's edge—probably checking for debris, if there's anything they can find there," Cpl. Vardy told CBC. "They'll still look around for any exit tracks, any possibility that he may have actually gotten out of it. They'll look around to make sure everything's okay there. We've also got a camera, an underwater camera, coming in today from the underwater recovery team in Newfoundland and we'll send that up and see if we can drop the camera down and see if they can pick up anything from the bottom."

The bottom of Makkovik Bay was 27 metres down through murky water beneath that hole in the ice.

"It's not looking good right now, I can say that," Vardy continued. "It's a great possibility, but there are chances. You know, I can't rule it out, but right now there's no sign of any exit tracks coming out of it. Right now we have tracks heading directly into the water, but nothing coming out ... Right now we're looking for a missing boy. We'll continue looking for the boy and hopefully with any luck he'll be found. Right now our concentration—everything's indicating to this hole in the ice and that's more than likely where he's going to be found."

Vardy wasn't the only person who thought that the hole marked the end of Burton's ordeal—in fact, it was becoming a widespread belief throughout the community and beyond.

"The worst fear was that he was sunk and went into the water," Andersen said. "That's what we told the family."

As it turned out, the underwater equipment was not

needed, or used. The RCMP plane that delivered it, however, became invaluable to the search effort.

"When it [the RCMP Pilatus] came in we went to the pilot and asked him to take us up to look around," Edmunds said. "When we went out there were five of us on board. On the second pass we spotted [Burton's Tundra 300]. We wanted the pilot to slow down. We couldn't determine if there were more tracks."

All that the searchers in the plane could do before landing was radio the detachment with the location of the snowmobile.

The timeline states: "RCMP plane arrives in Makkovik and offloads the equipment. RCMP in Makkovik ask the pilot to take on some searchers to fly over a hole in the ice to look for debris or tracks. The RCMP plane takes off. An abandoned snowmobile is spotted out on the sea ice during the search, remote from the town. Ground searchers were immediately dispatched to the location to follow up. They were not able to get to the snowmobile and it could not be determined if the driver was in the area. RCMP plane left due to impending darkness."

The discovery of the snowmobile prompted the RCMP to make their second request to FES-NL for some kind of search aircraft; FES-NL, in turn, made another call to the relevant military agency, as well as to their usual civilian contractor.

Meantime, C. Macdonald at JRCC Halifax had independently reopened the Burton Winters incident file almost an hour earlier after being asked by L Forces Northern D201 (the Canadian Rangers' land forces division called Canadian Forces Northern Area) about the procedure for providing helicopter support to look for the missing Makkovik teen. After explaining the procedure, Macdonald contacted Goose Bay

Ops. He was waiting for an update on the state of the 444 Squadron's two Griffons when FES-NL's Paul Peddle called again to officially request the vital air support.

In exchange for information—which turned out to be inaccurate—about what had been found, Macdonald could only offer Peddle the same bad news that he'd given FES-NL the day before.

"Paul Peddle EMO requesting air support to do search of region," the JRCC incident log reads. "Snow machine was found underwater and they believe it is possible to make it to shore from there. Explained we have no Herc at this time and our 412 in Goose Bay is US." Peddle called back shortly afterwards with a correction and a stronger plea for help: "Skidoo was found on top of the ice and they feel there is a realistic chance the boy could still be alive. They cannot search due to darkness. Advised him I am waiting to hear back from my OIC and will let him know."

Macdonald heard back from his officer in charge in about 10 minutes.

"Discussed options," he wrote in the log. "He wants 444 to go if they are serv in next hour or so and if not send Aurora. He does not want to send the Corm [Cormorants] with no serv Herc in the region."

At the time of the search, the Royal Canadian Air Force had five Cormorant helicopters and four Hercules transport planes stationed within range of the search area in northern Labrador. The Augusta Westland CH-149 Cormorant is a 22.81-metre air-sea rescue helicopter with room for 45 standing soldiers and a crew of five: a flight commander, a first officer, a flight engineer, and two search and rescue technicians. An empty Cormorant weighs 10,500 kilograms and can fly up to 1,389 kilometres. DND had four Cormorants stationed with the 103 Search and Rescue Squadron in Gander and one at 14 Wing Greenwood. Three of the four Gander Cormorants were in working order and the Greenwood aircraft was then on 24-hour standby and ready to fly at almost a moment's notice. The Lockheed C-130 Hercules is a four-engine turboprop military transport with a crew of five. It has a flying range of 3,800 kilometres. The Hercules is sometimes used for search and rescue operations, but it is usually only equipped with weather and navigational radar.

All three of the Hercules the RCAF had stationed at 14 Wing Greenwood were out of service at the end of January, but one Hercules was ready to fly out of a DND base at Trenton, Ontario.

The 444 Squadron informed Macdonald that one Griffon's leaky oil lines would be repaired within the hour (by 6 p.m.), and he recorded that the 444's one serviceable Griffon could be officially tasked for the Makkovik mission.

That Griffon helicopter left Goose Bay at 7:38 p.m. and arrived in Makkovik more than an hour afterwards.

In the meantime, Macdonald spoke with Vardy to get more information about the object of the search, what had been found so far, and the current conditions around Makkovik.

"He feels the 14 YOM may have been either lost trying to go to Goose Bay," the log reads. "He would like us to search Cape Strawberry and Ford's Bight toward Makkovik."

A new controller took over at this point, Captain Kristin Macdonald, who also spoke with the RCMP officer, transcribing his questions and Vardy's answers in detail and entered them into the log. The conversation went something like:

"Where were the tracks coming from?" Macdonald asked.

"The tracks came out of Makkovik Bay and towards

Strawberry Point," Vardy answered. "The snow machine is pointed towards Cape Strawberry."

"Are the tracks still visible or has it snowed?"

"It has snowed eight inches since the incident."

"Would it have been logical for him to walk back to Makkovik via Ford's Bight or via Makkovik Bay?"

"The shortest way would have been via Makkovik Bay."

"Is the snow machine still at the incident position?"

"Yes, GSAR has not been able to get to it."

"Is it broke through the ice?"

"No, it looks to be on top of the ice. However, there is a 60-foot open stretch of water and we are not sure if he went into that water or made it off onto safe ice."

The discovery of Burton's snowmobile had changed the nature of the search. As Edmunds explained, "Right then the focus shifted from the water to the ice beyond the Ski-Doo. The boys [one of the ground search and rescue teams] went out along the shore, chopping foot holes in the ice so they could climb up onto the ridges to where they could look around for him."

The ground search party nearest the snowmobile did not find Burton before darkness began to fall. Even as Vardy was telling Macdonald what was happening, that nine-person search team had to walk back off the ice to where they had left their own snowmobiles in a shallow inlet on Cape Strawberry. Vardy told Macdonald that all the searchers would soon be driving back to Makkovik for the night.

That did not end Tuesday's search efforts. When the 444's Griffon arrived in Makkovik at 8:45 p.m., it almost immediately took back to the air to start searching over the sea ice, following a series of instructions issued from the JRCC in Halifax: "Search a 1-mile radius around the incident position, then conduct a shore crawl from Cape Strawberry South down the West shore until the end of Ford's Bight. Then back out the West shore of Ford's Bight. Once Ford Bight is complete return to the incident position and complete a track crawl South West to Makkovik Bay towards Big Island and into the town of Makkovik. That will probably use up your fuel."

About halfway through its first sortie, after it flew over the location of Burton's snowmobile, the Griffon reported in to the JRCC, but its crew had only one new detail to offer: "Searched LKP [last known position]. Gas can on the ice behind snowmobile. We are searching down Ford's Bight now."

During that conversation, Wild Bight was added to the Griffon's search objectives, as requested by the incident commander. Consequently, the 444 Squadron's helicopter stayed airborne a little longer than had originally been planned, not landing to refuel until after 10 p.m. At that time its crew gave a more detailed report of its activities during the flight: "Fueling est 30-40 minutes. Will call back before T/O [take off]. Checked out co-ordinates passed by RCMP on a bluff, negative. Checked area around sled and to east pretty thoroughly. Nil seen. Gas can was 5-10 ft behind sled, lots of cracks and there is a lead near the shore close to the sled. Cliffs are 20-30 feet high in that area ... Winds NW (north-west) 20-30 (knots), good visibility and about ½ moon."

The 444's Griffon finished refuelling and began its second sortie at 11:20 p.m. to re-examine the area it had covered earlier. The aircraft stayed out in the dark, scouring the surface of the ice, for almost two hours.

In the meantime, JRCC Halifax once again arranged to bring another DND aircraft into the area for that same night, despite the extremely pessimistic "best case scenario" it had generated with the Cold Exposure Survival Model for Burton's chances earlier in the evening: "Log states search object has been missing since Sun night. Assuming that means 29/0000 the search object has been missing 48.5 hrs. Survival time > 36 hours. Functional time 29.2."

That evening, a Canadian Forces Aurora 113, usually based at 14 Wing Greenwood in Nova Scotia, was engaged in a training mission over the Strait of Belle Isle near the Labrador border, only a few hundred kilometres south of the search zone. The JRCC asked that the Aurora, designed primarily for coastal surveillance and anti-submarine warfare, be immediately retasked because of its Electro-Optical/Infrared (EO/IR) suite, highly advanced technology that helps its crew see in the dark. The four-engine, fixed-wing CP-140 Aurora, which has a flying range of 9,300 kilometres, is also equipped with sonobouys, radar, a Magnetic Anomaly Detector, an Applanix DSS-439 Digital Mapping Camera, and a pair of gyro-stabilized binoculars. After a quick flurry of official requests, the Aurora got its new orders and changed course to head north a few minutes before 11 p.m.

"JRCC contacted 14 Wing Greenwood (Nova Scotia) Operations to commence work to redirect a Canadian Forces Aurora from training mission to support the Makkovik search and rescue"—this is how FES-NL described the process. It continued: "The Aurora from Greenwood re-tasked to assist. As a secondary SAR asset, it was re-tasked to respond using its night search capacity with its Electro-Optical/Infrared (EO/IR) suite … Aurora arrives in Makkovik with sufficient fuel to provide approximately 1-1.5 hours of search time."

When the JRCC dispatcher called the 444 Squadron's officer in charge to relay the Aurora's estimated time of arrival at the search zone, Captain Gillis gave the centre another report about the Griffon's first sortie: "Had a good look at the LKP and spent a lot of time on the North West side," Gillis told the dispatcher. "Also completed Ford's Bight and Wild Bight. There was an area that the incident commander wanted checked out where a lot of crows were seen. There was absolutely nothing there. The visibility is excellent. We saw the snowmobile 1 nm [nautical mile] back."

The Aurora first flew over the scene at 11:42 p.m. The plane had enough fuel on board to remain in the area only 90 minutes, so it immediately commenced flying its designated search grid around the nearby bays and over the icefields—the EO/IR equipment sweeping the cold expanse for a tiny point of heat.

As the third day of the search for Burton Winters ended, two Canadian military aircraft were in the air east of Makkovik trying to locate a youth on foot, and dozens of local and out-of-town searchers were awaiting the return of a Universal Helicopters chopper, which FES-NL had contracted and arranged to have fly from Goose Bay to Makkovik first thing in the morning, possibly bringing with it trained spotters—highly experienced members of the Civil Air Search and Rescue Association (CASARA). At least, that was the hope of the JRCC controller and the advice he gave to FES-NL.

"Passed contact info for CASARA Goose Bay" to FES-NL's Paul Peddle, the JRCC controller wrote in the log. "Highly recommended they be used on the aircraft tomorrow. Mr. Peddle has never heard of CASARA. EMO [FES-NL] agreed that they would have chartered aircraft available to resume air search for first light."

Day four: Wednesday, February 1

At 12:10 a.m. on February 1, the latest known weather for that area of Labrador's north coast was entered into the JRCC log: "High scattered [clouds], visibility unlimited, temperature minus 16."

Both military aircraft had to depart from the expanding search zone at around 1 a.m. The Griffon was the first to contact the JRCC, and the crew reported a hopeful new discovery: a sign of life.

"We searched area again and had a good look at the rocks to the NW of the LKP," the crew reported, as recorded by the controller. "We went back to the sled to have another look and the FE [flight engineer] noticed footprints leading away to the South of the LKP straight toward land. The tracks were only visible for 150 feet. Then we lost the trail. They were definitely heading to the shore. The GSAR [ground search and rescue] team may be able to pick them up."

The find was reported on the FES-NL timeline: "Griffon completes their search area twice. Griffon Flight Engineer makes first discovery of tracks from snowmobile. Footprints lead away from the South of the Last Known Position straight toward land. The tracks were only visible for 150 feet."

The sight of those footprints leading away from the snowmobile added to the hope sparked by the discovery of the machine, and it galvanized what might otherwise have become a flagging search effort. The footprints provided refreshed confidence to those who had told themselves to expect the worst possible outcome when they looked down the icy hole into Makkovik Bay and probed the frigid waters for a big machine, or a small, still body.

"We never gave up," Vardy said. "We continued to look

along the water's edge. We deployed teams in different areas. You know, there's always hope and we never give up on hope."

For a short time that morning, Vardy explained, the searchers felt enough hope to wipe out all their previous pessimism: "Today ... when we found the track today leading away, you know, everybody's heart jumped for joy and everybody was excited and everybody was eager and trying their best to bring this to a positive close."

The Griffon and Aurora flew as long as they could, but neither aircraft found anything else of significance during those dark early morning hours. The Griffon was sent back to the Goose Bay air base at 1:38 a.m., arriving 80 minutes later, even though the local RCMP officer in charge earlier expressed a strong desire for the aircraft to remain on the coast overnight.

"Cpl. Vardy would like for Griffon to spend the night in Makkovik and go back out tomorrow," Macdonald recorded in the JRCC log during the afternoon. "I advised that AC will return to Goose as it is a fairly short transit and the aircraft commander is concerned about aircraft security. Also he was informed that the Griffon only has one crew and they would be on crew rest until 1700z. I suggested that he make arrangements with EMO now so they could have full day tomorrow. Cpl. Vardy said EMO would not entertain idea until night search was complete. I said that I will call EMO. The weather will be very favourable tomorrow and EMO asset should be ready for first light."

That EMO asset—the Universal Helicopter aircraft chartered by FES-NL—didn't leave Happy Valley-Goose Bay until almost 8:30 a.m.; it arrived in Makkovik at 10:15 a.m. Ground teams had gathered with their snowmobiles at the RCMP detachment building at 8 a.m. to muster for the morning search and they immediately went back to the shoreline where the footprints had been spotted by the Griffon. Despite their intensive efforts, the ground searchers did not find any signs of the teen on the surface of the ice.

When the aircraft from Universal Helicopters landed in Makkovik, Randy Edmunds, Barry Andersen, and one other man got on board to act as spotters. They flew to the ice ridge where Burton had abandoned his machine and they picked up the trail of his footprints. With only the use of their eyes and some ordinary binoculars, they tried to retrace the route the teen had taken toward the mainland.

"I was in the front seat," Andersen recalled. "The guy [the pilot] was following the tracks. We knew it was probably him. We were not in a recovery. We were still in a search operation."

Edmunds, who was in one of the back seats, described the final minutes of the search: "On Wednesday the chopper came in from Universal Helicopters. It took us right to the Ski-Doo and then we started tracking the footprints. We were up about 100 feet and it took us one and a half hours to go the 19 kilometres."

From what Edmunds could see of the footprints, it appeared that it had taken a long time for Burton to tire. At first the prints looked firm and steady, but eventually they indicated that he had become exhausted. Edmunds's summation: "He was walking continuously ... at the end it didn't look like footprints, it looked like skiing. He was dragging his feet. He didn't have any energy left."

Then, suddenly, after looking for Burton Winters for almost four days, the searchers saw him far below.

"I've found dead people before, but that doesn't make it easier." Edmunds thought he saw a tiny black spot in the snow, but he hesitated to tell the others in case his eyes were playing tricks on him—maybe they were showing him what he wanted to see and not what was really there. He waited for someone else in the helicopter to independently confirm what he thought he had spotted.

"I waited for Barry to say he saw something black on the ice—and then he did," Edmunds stated.

The team was supposed to report the teen's location for someone else to retrieve him, but they didn't want to fly away and leave him there. There was always the chance a ground team would not be able to reach him by travelling on the surface.

"The chopper pilot said we've got to call search and rescue to pick him up," Edmunds noted, "but we made him fly over the water to see if there was enough ice to land on."

They convinced the Universal pilot that the ice was thick enough to hold the helicopter and he landed it near the teen.

"[We] went to pick [Burton] up," Edmunds recounted with some emotion. "He was frozen solid except for a buckle around his abdomen. We managed to pick him up in our arms and we walked back to the helicopter and laid him across us inside."

The helicopter, with the searchers and Burton's body on board, flew back to Makkovik within minutes. Burton was taken from the landing site and driven to the local nursing station for the medical staff to perform a desperate attempt at reviving the severely hypothermic teen. The Makkovik Community Clinic staff are not permitted to publicly discuss what happened, but a report written later to assess medical transport and respiratory therapy delivered to Labrador's north coast notes that a 14-year-old, who had been missing for three days, was brought into the clinic at noon on February 1, when the temperature outside was -37°C.

"Clinic staff have started CPR," the report reads. It doesn't say how long they persisted at their efforts, but others, including Burton's mother, said they didn't give up for a full four hours.

Reflecting on the search, Barry Andersen could find only one small consolation: "At least we found the body to bring some closure for the family and the rest of the community as a whole. I guess not only the community here in Makkovik, but Postville, Hopedale, Nain—he has relations all up and down the coast here."

The JRCC Halifax log does not declare the case closed again until four days later. It took that long because the JRCC first received inaccurate map coordinates for where the body had been located. They were unwilling to close the file until they had received the correct information from the RCMP.

part three
One small boy, one big national debate

1. Chinese lanterns

It takes at least two people to light a Chinese lantern and set it aloft. One person could try it alone, but the paper envelope might flop down over the wire frame and touch the little flame in the middle. To do it safely, one person—preferably two—should hold on to the paper, opening it up and giving it form, while another lights a match and sets it to the wick. As the paper envelope fills with warm air, it slowly expands until the whole contraption becomes light enough to float by itself. When the person holding it lets go, the lantern will hover for a moment, then gently rise up into the night sky.

Late in the winter of 2012, in many places around Newfoundland and Labrador, dozens of Chinese lanterns shone in the night sky after word had spread that the body of the 14-year-old from Makkovik had been found. These lanterns were all set aloft for one purpose: to give Burton Winters a light to follow off the frozen sea, to give him hope, and to show him the way home.

That's what the candles were for, too, as they were set in windows all over the province or carried in solemn nighttime marches and carefully placed in the snow. So were the headlights of cars and snowmobiles switched to high beam and pointed at Makkovik, no matter that the town was hundreds of kilometres away.

No matter, either, that it was too late. After three days of waiting to hear that Burton had been found alive and well, even after there was no reasonable hope that such news would come, few were ready to accept that he was gone. Many acted as if there was still some part of him out there somewhere, trying to find his way off the ice, some part that might see the lights and know he had not been forgotten. People wanted to show him that they had not given up but were doing everything they could to help him. Adding more grief to Burton's family's sorrow, and injecting bitterness into those vigils, was the growing realization that maybe everyone did *not* do everything possible to find the youth lost in the middle of a fierce Labrador winter.

Beginning immediately after the tragedy, and continuing for weeks afterwards, nighttime vigils were held in many Newfoundland and Labrador communities, involving thousands of people who had never met Burton Winters or even heard of him before he went missing. No matter. In Makkovik, Postville, Hopedale, Happy Valley-Goose Bay, North West River, Labrador City, L'Anse au Loup, Port Hope Simpson, Cartwright, St. Anthony, St. John's, Sheshatshiu, Natuashish, Pilley's Island, Englee, Eastport, and other places, groups numbering a few dozen to several hundred came together to share their emotions with each other and with the governments in Ottawa and St. John's.

Burton's family were involved in the vigils, as participants and organizers. A week after his son was found, Rodney Jacque stood with more than 100 of his fellow townspeople on the ice of Makkovik Harbour and addressed the crowd. "You can't change the past," he concluded. "You can't change what happened. You can't change anyone's mistakes. That's done. But we've got to change things for the future so that this doesn't happen again, so no one else has to go through this. It shouldn't happen in this day and age."

Those at the vigil expressed astonishment at how far Burton had walked after abandoning his snowmobile and how hard he had tried to find his way off the shifting ice. It struck the searchers, too, as they followed the trail of his prints through the snow.

"The first few kilometres where it was in the drift ice and the ice pans we had to, you know, we'd see a footprint and then we'd lose it," Randy Edmunds recalled. "We'd backtrack. We'd take an educated guess. It was just zigzagging around and, after a while, we were looking at each other and saying, 'He's still walking!'"

As heartbreaking as the outcome was, the effort Burton had made to return home made his family proud of him—not that they weren't already.

"[It is] very hard to measure what he went through those last few hours, those last moments," said his stepmother, Natalie Jacque. "He's so strong, very strong. When we heard the distance, for me it showed how much of a man he was and how much of a man he could have been. I just think about him walking, trying to get home and just not wanting to give up. Every night when I try to go to sleep that's all I can think about: my little boy walking on the ice."

That spirit provided the basis for a nation-wide "He Walked This Far" campaign. As Eastport Mayor Genevieve Squire explained during a candle and flashlight procession in early March 2012, "We were dressed warmly and we only walked 1 kilometre, over paved roads that were plowed. Burton walked 19 kilometres over sea ice in freezing temperatures before he died. We walked in his honour. It's hard to imagine what he must have gone through."

The slogan "He Walked This Far" appeared on signs held during vigils and became the titles of songs written in memory of Burton. The most popular of these songs was composed, performed, and recorded by a young Straits woman, Jamiee Thomas. "I hope that it will make people push for more help that we need in Labrador for search and rescue," Thomas said about her composition.

It wasn't long before the vigils became protests. While many of the placards continued to display expressions of grief, more and more of them became accusations and demands. At first they contained such sentiments as "You'll always be in our hearts, Burton," but soon questions such as "Why leave when you have lives to save?" or "What if it was your child?" or "How can MacKay say DND's response was satisfactory?" joined them, along with others that read "One death too many" and "One call should be enough!" Demands were also emphasized: "Independent inquiry!," "Primary SAR in Labrador!," and "Keep SAR in NL, Burton Rest in Peace."

Some were heartbreaking: "My big cousin should have been rescued ... now he's just a memory." The son of Burton's aunt Stephanie Fost held a placard bearing these words in St. John's during one of the first vigils in front of Confederation Building. Fost spoke to the 30 or so who gathered there that night: "Newfoundlanders and Labradorians need to take a stand. We need to take a stand to honour Burton's legacy and to make sure no other lives are lost."

Fost had set up that vigil after learning of the ones held in Labrador. "As we continue to grieve over the death of our 14-year-old Burton, we are speaking up today and will continue to speak up and hope that a search and rescue unit will be stationed in Labrador that will be adequately staffed," she said. "We understand that Burton will never come home. However, we would like future Burtons to have a fighting chance to come home to their loving friends and family. In honour of Burton Winters, we are pleading for change."

When Burton's family first held a vigil for the youth, it was not about politics—only hope and fear and the comfort their friends and supporters offered as they waited for news that the searchers would bring back from the woods, hills, and frozen bay. That vigil began the night Burton went missing and continued until his frozen body was carried back to town. Many family members who were not in Makkovik travelled there as soon as they could; others held vigils wherever they were.

Charlotte Winters-Fost, Natalie Jacque's mother, was known to Burton as "Nanny Charlotte." Although she lives on the island of Newfoundland, she was in Calgary, Alberta, to attend a grandchild's christening when her daughter's stepson was reported missing. "For the first two or three nights that he was out, knowing that he was out in the cold and that I was going to bed in my own warm, comfortable bed, it was really difficult knowing that he was out there all alone in the middle of the night, in the middle of the winter on the ice in northern Labrador," she lamented.

Burton's father tried to describe how he felt about Burton's disappearance: "I don't know how to explain it. You're so worried you don't know what to think, just trying to think anything you can."

"We just hoped he was safe and would be home soon," added his wife. "I guess we were still trying to stay positive. We didn't want to think the worst. We didn't want anything to be wrong with Burton."

Burton's mother, Paulette Winters-Rice, was at her home in Happy Valley-Goose Bay when she heard about her son. Her aunt, who lived nearby, found out about the search online and shared the news with her. "At the time we were not even sure it was my son," Burton's mother explained. "So I called my mom and she was trying to find out if it was Burton." Winters-Rice flew to Makkovik the next day. By then the community's children were also holding a vigil for their missing classmate. Classes had been cancelled at J.C. Erhardt Memorial School on Monday morning, but most of the children went to school anyway.

"It sort of knocked everybody, students and staff, to our knees," admitted the principal, Elizabeth Mitchell. "Those days were like a dream. We were all hoping and praying he would be found and brought back to us."

Burton's cousin and best friend, Willie Flowers, spent most of Sunday evening at home, but he didn't remain there on Monday: "School was still on, but we didn't have any classes. We just tried to stay calm. The first day or two it was just teachers and they let us play. Then people from out of town came in to talk to us. It helped."

The children's vigil, like that held by their elders, came to a sad end, and no amount of counselling could soften the blow. "It was kind of hard to believe," remembered Jacqueline Winters, who was 13 at the time. "But after three days everyone lost hope. Everyone knew he was dead beforehand and everyone was miserable."

"We knew that day would come and when we heard it, it was like, okay, it's happened," echoed Dalton Manak. "Right after it happened school closed down. Everyone was talking together. We stuck together."

During those first days, some people couldn't allow themselves to imagine the worst possible outcome for even an instant. Winters-Rice recalled her reaction when two RCMP officers told her that her son's snowmobile had been found. "I was right excited, saying, 'Oh good! They're going to have him.'" But then the police returned with more news. "They were looking at me and not saying anything. Finally they opened their mouth and said they found him and I was happy. Then they told me it wasn't good and told me he was gone. After hearing that I fell to the floor. I fell to the floor crying. I could not even breathe. I told them to stop lying to me. I screamed at them to 'Stop lying to me! Stop lying to me!'"

The officers brought Winters-Rice to the Makkovik Community Clinic, where medical staff were attempting to revive Burton. "We were told they were doing CPR on him for over four hours and they told us they could keep going. But there was no point. He was not going to come back," she asserted.

The vigil over, their hope gone, the large, widespread family of Burton Winters turned to private mourning—but not for long.

Speaking about what happened in those first few hours after he had helped bring Burton's body back to Makkovik, Randy Edmunds indicated that there was hardly a chance to mourn before the media had started asking questions: "The next day I started to get calls from officials and from the media. I said, 'This is going to go crazy.' I told the family what I thought was going to happen and they said, 'We do, too.'"

The Jacque and Winters family immediately accepted the need to have their views and demands publicized as widely as possible. As Winters-Rice told the crowd on the ice of Makkovik Harbour a week after she had seen her son's body: "I live in Goose Bay and I want to know why we were able to get here before the rescue team." She targeted the helicopter DND eventually sent. "I want to know why and I want to know why the rest of our family was able to get here before the rescue team and, on topic, since we've been here the weather was really nice and everybody here played their part. Why didn't they do theirs?"

•••

The public movement to wrest something positive out of the terrible tragedy quickly gained momentum and led to the creation of an online discussion forum, a Facebook group called The Burton Winters' Rescue Center IN Labrador. This group, which attracted more than 36,000 members (more than the total population of Labrador), became a focal point for the public; it was used not only to lobby government for the SAR centre but also to plan actions and exchange condolences and memories.

"The goal of this group is not to express our political beliefs other than to advocate for change; so this never happens again," founding member Adrienne Morris wrote as the group's statement of purpose. "Our goal is to get a SAR Centre in Labrador ... We are uniting here, the one place we can meet where all of our differences melt away and we can stand together in a purpose that unites us ... We welcome all posts that reflect the goal of this group to get a SAR Centre in Labrador."

The Facebook page has also been used for more sombre messages, like those marking important anniversaries. It was filled with memories and wishes for Burton on his first birthday after his death. "Burton, happy heavenly birthday!" one relation, Barry Lyall, posted. "Just saw your Grandmother Charlotte, Great Grandmother Winters, Aunt Ruth, Aunt Stephanie and Cousin Freddie and I have to say that's what I call a family that had great love for you! You were one lucky young man! I just lit a candle for you, young man. I will never give up on find out the answers to why you weren't rescued and getting justice through an inquiry for you!" "Happy birthday, Burton," wrote another. And another: "Hates to do it this way. Wishing it was in person, but things are not that way. But we will meet again—may angels lead you in." And, "Happy 15th birthday to my cousin Burton!" Nichole Jenkins posted. "I hope you are having a wonderful birthday up there! I wish you were here to celebrate. I love you and I miss you so much … one day we will meet again."

Many more followed. Most contained simple wishes that Burton rest in peace, but some contained memories. "Auntie's little one, you always came home with a smile and a drawing or homework in your hand," posted one of Burton's cousins, Marion Winters. "Sometimes you were crying and you would run into my arms. I would hold you until your tears stopped falling. I always said, 'I love you, baby, and no one is ever gonna hurt you. Just stay by me.' He always came to me … He was so grateful I was there for him.… Always remembered and never forgotten."

Facebook was used to organize letter-writing campaigns, petitions, and protests. One form letter was addressed to the prime minister: "Mr. Harper, I would really like to hear your response to this. Our country needs your support to help make Burton's death not be in vain." Another, addressed to Newfoundland and Labrador Premier Kathy Dunderdale: "We, the voting public and residents of Newfoundland and Labrador demand an independent inquiry into the tragic death of Burton Winters. We are outraged that both governments have failed to address this issue in the manner it deserves. We feel that none of us as citizens will be able to sleep soundly until the government holds those accountable for the incidents that led to the tragic outcome of this situation; along with the many tragedies that occurred before this one ... We demand that an inquiry be held and we look forward with anticipation for this to take place expeditiously."

Plans for demonstrations were often specific, as illustrated by this directive issued before a protest planned in Happy Valley-Goose Bay for March 28, 2012: "We are asking everyone to meet at the arena parking and we will march to MP Peter Penashue's office. We want three items addressed: 1. Independent inquiry; 2. Is the current equipment serviceable [?] and we want SAR in Labrador; 3. An apology to the family and the Province of Newfoundland and Labrador."

As passionate, organized, focused, and persistent as all the vigils, protests, and demands were, they did not achieve their goals. Neither the federal nor provincial governments undertook any real action to fulfill them.

By the February 9 vigil in Makkovik—during which Burton's mother and other family members voiced their heartbreaking questions—the authorities they wanted answers from seemed to have already considered the matter closed. In fact, just two days after Burton's body was found, DND held a press conference to present the preliminary results of what it called an exhaustive and conclusive investigation into its own conduct.

2. First responses

Burton Winters was first mentioned, although not by name, in the House of Commons in Ottawa the day after his body was recovered from the ice. The family had asked the RCMP not to release his name to the media, but it was not long before it became common knowledge. By the time Jack Harris, MP for St. John's East and the Opposition New Democratic Party's justice critic, raised his concerns about the Conservative government's search and rescue policy in Question Period, almost everyone in the country knew who he was talking about.

Harris's first question, however, was not about Burton but Newfoundland's doomed Marine Rescue Sub-Centre: "The Conservatives are continuing their shameful plan to close down the marine search and rescue coordination in St. John's. The government is putting lives at risk by closing a vital search and rescue facility along one of the world's most dangerous coastlines. The expertise of the St. John's search and rescue centre is vital to ensuring the safety of Newfoundlanders and Labradorians and all Canadians. Why do the Conservatives have billions in tax breaks for profitable corporations, but no money left to keep open this vital search and rescue centre?"

Keith Ashfield, the Minister of Fisheries and Oceans, responded: "Mr. Speaker, I have answered this question many times. We will not put in jeopardy the lives of our mariners. We will not do that. This has nothing to do with that. We have an opportunity to make consolidations to save money for the tax payers of this country without putting anyone in jeopardy or at risk."

Harris was not satisfied. "Mr. Speaker, the body of a 14-year-old boy, missing since Sunday on the coast of Labrador, was found yesterday, but search and rescue helicopters did not arrive until Tuesday night," he began, but then questioned, "Why is it that the Conservative government gives such a low priority to search and rescue? With our response times well behind international standards, the government is closing down rescue coordinating centres and helicopters take two days to start searching for a lost boy." And then he added: "How can a search and rescue helicopter be available to transport the Minister of National Defence, but not be ready to search for a lost teenaged boy on the coast of Labrador?"

Harris drew two comparisons in his one public statement. First, he targeted the government's intention to close down the Maritime Rescue Sub-Centre in St. John's. The government's rationale was that closing the sub-centre would save money by consolidating all of the organization's administrative and operational functions in a central location—in this case, the Maritimes. Opponents of this closure argued that the sub-centre provided a vital service to Newfoundland and Labrador mariners and lives should not be put at risk for the sake of a budget-cutting measure. Although the sub-centre was not actually involved in the search for Burton Winters, its closure was frequently cited as another example of the Conservative government's neglect of the country's search and rescue capabilities. Second, Harris referred to an often-cited example of the government's active abuse of the system. In 2010, Defence Minister Peter MacKay had been able to call, on short notice, one of Gander's Cormorant helicopters from its normal duties to pick him up at a fishing camp and ferry him to the nearest airport so he wouldn't be late for a Conservative Party function in southern Ontario. This action was declared a misuse of public resources; in light of it, that same helicopter's absence from the Makkovik-area search was regarded as even more reprehensible and heartless.

"I was shocked. I think we all were," Harris summed up after his statement in Question Period. "There's something very wrong here. Here we are six months after people were scandalized by the use of the transport of Peter MacKay as defence minister in a search and rescue helicopter and yet they're not at the ready when it comes to this kind of search."

That day in the House of Commons, the Conservative government did not respond to either point.

"Mr. Speaker, certainly the death of this young man is a tragedy and our condolences are extended to his family and friends," responded associate Minister of Defence Julian Fantino. "I have asked the officials to look into the incident. I can inform the House that the Chief of Defence Staff has commenced an investigation. Search and rescue teams work with federal, provincial, and municipal partners to respond as quickly as they possibly can to save the lives of those at risk. Search and rescue crews react as quickly as possible every time. We will get the answer and it is forthcoming."

"We want to see that investigation happen very quickly," Harris remarked afterwards. "And we'd also like to see the details made public."

In St. John's, Merv Wiseman, a long-time Coast Guard employee and the union representative at the Marine Rescue Sub-Centre, had already gone public with his concerns about federal cuts to the search and rescue system. Wiseman, who had run as a Conservative candidate in the riding of St. John's-Mount Pearl in the 2008 federal election, turned against the government he had wanted to join and declared Stephen Harper's policies dangerous to human life. For months, Wiseman had warned that eliminating both the St. John's sub-centre, where he worked as a coordinator, and the one in Quebec City, would hurt the ability of the agency to find and rescue people in distress.

In July 2011, Lieutenant-Colonel John Blakeley, spokesperson for the Canadian Forces, claimed that the St. John's and Quebec City facilities could be shut down safely because decentralized sub-centres were no longer necessary: "We've just reached a point where technology allows us to do everything out of the three main joint rescue coordination centres." Blakeley added that the Coast Guard cuts contributed to meeting the Department of Fisheries and Oceans' $56 million reduction in its annual budget.

Wiseman countered by arguing that eliminating the local sub-centre would also eliminate the local knowledge that helped staff fulfill their duties: "The rescue coordinators in Halifax and in JRCC Trenton or JRCC Victoria have the same basic qualifications that all of us would have in St. John's. The difference is in the local nuances and the accumulation of local knowledge that we've acquired over the years ... Seconds count. This is where we lose effectiveness, as we move farther away from our shores."

When the Canadian Coast Guard revealed on February 1, 2012—the same day Burton's body was found—that it would accelerate the closure of the two rescue sub-centres, Wiseman issued a statement as the Union of Canadian Transport Employees (UCTE) union shop steward for Local 90915, although he did not mention the Makkovik incident explicitly: "Countless studies and inquiries concluded these sub-centres were essential because only they have the local knowledge and expertise that saves lives. None of that has changed."

What *had* changed from that day to the next was the public interest in Canada's search and rescue system. Because of intensive news coverage on radio, on television, in print, and over the Internet, thousands of people—likely tens of thousands—knew and cared about the search for the lost youth.

Wiseman spoke up again on February 2, 2012, this time

to reflect on the search for Burton Winters: "What happened between Sunday evening and Tuesday evening is beyond me. It doesn't make a whole lot of sense at this stage."

The growing conviction that more could have been done to save Burton sharpened grief and evoked deep anger among the family members and the public. Even the federal government realized it couldn't entirely ignore the demand for answers.

Ottawa detailed Rear Admiral David Gardam to deliver some findings of the federal investigation at a press conference in Halifax on February 3, the Friday after Burton's body was found. By then, the question first asked by Burton's family was being echoed across the country: Why did the Canadian military not send a helicopter to help in the search until it was too late?

"Given the weather conditions, which were below limits for safe operations of an aircraft, our aircraft were not able to operate in that environment … We have to manage a very large area and it's a balancing act on how you manage weather, resources, aircraft availability, crew rest. It is very much like a ballet and it has to be managed that way," Gardam explained. He also said that DND considered its role to be secondary in this emergency, and the agency did not feel its services had been required. "Just so it's clear, under this type of search and rescue, we are called to provide services if civilian aircraft cannot be used," he told reporters. "Civilian aircraft were capable of flying during the day when the weather improved and there was no subsequent request for us."

DND likely hoped that Gardam's explanation would satisfy Burton's family and a growing number of others who were keenly interested in what had happened, and why. DND would have been disappointed—instead of allaying criticism,

Gardam, as commander of Joint Task Force Atlantic, instantly became its focal point. People knew that there were more officials involved than this single naval officer, but in Gardam they had the man who said he'd actually made decisions during the crisis.

Although Harris spoke for the NDP, he voiced a concern that many were feeling about the facts as presented. "It's clearly very preliminary," he told the Canadian Press. "The timeline is very sketchy and raises more questions or as many questions as it answers in terms of why a helicopter or assets of the Canadian Forces weren't available more quickly. I'm not satisfied that we know the answer to that question yet and we want more information. And we want a fuller report and a fuller investigation."

Even before Gardam's press conference, northern Labrador's Nunatsiavut Government had already prepared a scathing denunciation of DND's lacklustre participation in the search. Gardam's statements at the Halifax press conference did nothing to change their position. The Nunatsiavut Government demanded that Ottawa establish a permanent search and rescue station in Labrador.

"There are many questions that may never be answered as to why this tragedy occurred, but we have to ensure something like this never happens again," was the statement from president Jim Lyall. "We truly believe that Burton would still be with us today if the search and rescue response time had been quicker … We understand the calls for search and rescue were made shortly after the boy went missing, but the air support out of Gander and Goose Bay were not available. That is totally unacceptable."

Politicians weren't the only ones who responded publicly to the military press conference. The Happy Valley-Goose

Bay CBC radio station broadcast numerous statements from north coast residents.

"I'm very disappointed on what happened to search and rescue for Labrador," said Dan Michelin of Rigolet, an Inuit community on the narrows between Lake Melville and Groswater Bay. "Every community in Labrador should get a petition going and send it to the federal Minister of Defence, stating that we need a substation here in Labrador. If Gander's only for offshore Newfoundland, then we should have one for Labrador, as well."

Toby Anderson, calling from Makkovik, spoke about a successful search operation that had taken place six months before: "The rescue by search and rescue of three men from the cliff in Cape Harrison [one of the most dangerous capes on the coast of Labrador] this past summer showed us how effective the search and rescue service can be. The Burton Winters tragedy was the opposite. The excuse given by the military that they couldn't deploy the appropriate aircraft from whether it was Greenwood or elsewhere because it might be needed somewhere else tells us that the life of a 14-year-old Inuit boy is second to anybody else."

Some calls were personal attacks. "Gardam's excuse for him not doing his job is to say that he didn't deploy the helicopter because there was already people on the scene and he needed a second call," said Jodie Lane, also a Makkovik resident. "That's a pathetic excuse, Rear Admiral Gardam, and not only should the people demand that more investigation be done into this to see what changes need to be made in search and rescue, but I also think there should be a performance review done on Rear Admiral Gardam to see if he should be deserving of keeping his job."

Harsh criticism came, too, from Rodney and Natalie Jacque. In a written statement, they decried "poor decision-making" and demanded more information about "someone else's failure to do their job." They further expounded: "This is not a time for excuses! This is the time for someone to step up and take responsibility ... to ensure that this doesn't happen to another innocent family. How is it that a civilian helicopter arrived on the scene, yet a Search and Rescue helicopter did not?"

The Jacques stated that the Canadian Forces should not have expected the aircraft volunteered by Woodward's Oil to adequately fill in for the military, since "the civilian helicopter which first arrived was neither equipped nor capable for a search and rescue situation" and the pilot "only offered to help because Search and Rescue had not yet arrived." Geoff Goodyear, CEO of Universal Helicopters, told a reporter that he didn't want to contradict DND, but he did want to explain that weather did not prevent the company's aircraft from flying to take part in the search: "We were delayed by two or three hours before we could depart Goose Bay and then we got on site around 11:30 or midday on the 30th." Goodyear's statement convinced the Jacques they were right to question the military's explanations.

3. DND's self-examination

On February 7, Defence Minister Peter MacKay rose in the House of Commons to say that he had *just* called for an investigation, by DND, into its own actions. "With respect to when we will have more answers, I met with the Chief of Defence Staff this morning," MacKay said. "I have asked him to have a full investigation into all the circumstances around this tragic death. We should have answers this week."

The "full investigation" must have seemed extraordinarily quick, since MacKay was able to report it finished only 24 hours later.

"A full investigation has now been completed," MacKay told the House on February 8. "We have a much greater understanding of the timeline and the way that these tragic events unfolded. Both the RCMP and the Canadian Forces have explained some of these circumstances. There are improvements that can be made perhaps in protocol and we are in a constant state of update and improvement."

In actual fact, the investigation had been launched by a verbal directive from the Chief of Defence Staff on February 2, and its findings detailed in a report hand-dated February 7.

The report's stated mandate was "[t]o establish the facts of the incident and the Canadian Forces response ... To compare the Canadian Forces response to the established procedures and protocols ... To highlight any deviations from established practices ... To make findings on the appropriateness of actions undertaken by the Canadian Forces and ... To make, as needed, appropriate recommendations for further action." To conduct the investigation, the Strategic Joint Staff worked "in conjunction" with the Canadian Forces' Canada Command, the Royal Canadian Air Force, and 5 Canadian Ranger Patrol Group (the Makkovik Canadian Ranger patrol).

The DND report describes the beginning of the search operation: "On Sun 29 Jan 12 at 1900hrs Atlantic Standard time (AST) a 14-year-old male, Burton Winters, was reported missing from Makkovik, NL. He had last been seen approximately 5½ hours earlier (at 1330hrs AST 29 Jan 12) departing his Grandmother's house, alone and on snowmobile. The teenager was a Junior Canadian Ranger (JCR) and had been involved in a JCR outing earlier in the day, but he had returned from the outing prior to being reported missing. In accordance with normal procedure the Royal Canadian Mounted Police (RCMP) treated this as a 'missing persons' case and initiated a Ground Search and Rescue (GSAR) effort on Sun 29 Jan 12 at 1930hrs AST (6 hours after he was last seen). Of note, the local RCMP in Makkovik has the mandate to lead all GSAR efforts within its jurisdiction, including the decision to request CF assistance."

The report mentions the military's limited jurisdiction in the case but also highlights how extensively the Canadian Forces had been involved—mainly through their local Ranger presence. Having the Rangers on the ground may actually have contributed to DND's hesitation to become involved further: "From the initiation of the GSAR effort at 1930hrs AST on Sun 29 Jan 12, 10 Rangers and 2 Ranger Group Staff members from 5 Canadian Ranger Patrol Group (5 CRPG) assisted the RCMP with the GSAR effort. Initially they were acting as local volunteers, but a formal tasking to assist was issued at 0935hrs AST Mon 30 Jan 12. Throughout the operation, the Ranger Patrol Second-in-Command (2I/C) worked in the RCMP detachment headquarters in Makkovik assisting in coordinating the local rescue efforts as well as providing periodic updates to the Canadian Forces chain of command. Reports from the Patrol 2I/C during the day of Mon 30 Jan indicated that local weather conditions were poor and that aircraft were not able to conduct the search."

Poor weather is repeatedly cited by DND as the reason it did not send search aircraft, although it acknowledges the conditions were not severe enough to keep a non-military helicopter away: "The civilian aircraft engaged used different rules to determine safe weather operating limits for air search.

Universal Helicopters elected to fly in marginal conditions to Makkovik and then conducted air search operations, which were hampered by conditions that continually varied between suitable and unsuitable conditions."

According to the DND report, another reason—perhaps the main reason—that DND did not send aircraft was that a second call for help was not made. JRCC Halifax had requested a second call if the weather improved.

"This 'call back' procedure is the standard protocol followed by the JRCC and all provincial and territorial emergency management organizations," the report states. "This protocol is in place because of the high volume of both GSAR and maritime and aeronautical SAR events in Canada. By using the 'call back' procedure rather than having a continuing dialogue regarding all SAR events, each organization is able to focus on the specific events which require their direct attention and ignore those that are being handled by other agencies."

JRCC Halifax waited for a second call for the balance of the first full day of the search. When none came, the dispatcher, C. Macdonald, had no choice but to close the file. According to the report, "In this case FES-NL did not call back on Mon 30 Jan because they had successfully engaged contracted aviation assets for the daytime search. Approximately eight hours after the first notification of the incident, at 1718hrs on Mon 30 Jan 12, FES-NL still had not called back to request CF assistance. Consequently, in accordance with normal procedures JRCC closed the case."

The case was reopened late Tuesday afternoon, after the RCMP aircraft discovered Burton's snowmobile: "Later that same day (Tues 31 Jan 12) at 1643hrs JRCC Halifax received indications through the military chain of command that FES-

NL may submit a second request for assistance. The JRCC immediately re-opened the case and checked on the availability of CH146 Griffons in Goose Bay ... At 1654hrs AST JRCC Halifax received a second call from FES-NL indicating that the Burton Winters snowmobile had been found, weather had improved to permit search by aircraft, and that CF assistance was required to continue the search ... On Tues 31 Jan 12 the CF responded with a CH146 Griffon from 444 Sqn in Goose Bay, arriving on scene at 2045hrs AST equipped with night vision capability."

The report praises 444 Squadron: "It should be noted that the 444 Sqn CH146 Griffons in Goose Bay do not have a mandate to maintain a SAR readiness posture (i.e. 30 minute notice to move during the day or two hours notice to move at night), nor do they have a mandate to maintain a 'Ready 12' response time (i.e. 12 hours notice to move). Despite not having a mandated SAR response time, 444 Squadron responded extremely quickly by recalling the crew and getting Griffon airborne within two hours and five minutes of receiving the tasking; a significant achievement."

JRCC Halifax's action to get a second military aircraft to the search site is also mentioned: "While the CP140 Aurora is a secondary SAR asset it is equipped with an electro-optical/infrared camera system, which can assist in nighttime searches. Within 30 minutes the aircraft was retasked to the search, arriving on station at 2342hrs on Tues 31 Jan 12 and staying until it was forced to return to Greenwood due to fuel considerations."

Both aircraft left the north coast just after 1 a.m. on Wednesday, but by then the 444's Griffon had shown its value: "The Griffon Flight Engineer spotted human tracks in the snow leaving from the snowmobile. However, the aircraft was not able to follow these tracks beyond 150 feet from the snowmobile as the terrain presented by sea ice was harsh and varying and fresh snow obscured the tracks."

Despite that new clue, no additional aircraft was sent to Makkovik because no one asked for one: "During daylight hours on Wed 1 Feb 12 the GSAR effort continued using civilian helicopters; FES-NL did not request CF assistance with the daylight search on Wed 1 Jan [sic] 12."

The report outlines why the search around Makkovik was not technically within DND's jurisdiction: "Search and rescue services fall into three categories: ground, marine and aeronautical ... Under the NSP, Ground Search and Rescue (GSAR) is conducted under the legal authority of the individual provinces and territories. The RCMP is the operational authority for GSAR in parts of Newfoundland and Labrador. Fire and Emergency Services Newfoundland and Labrador (FES-NL) are called upon to assist the police forces (Royal Newfoundland Constabulary and Royal Canadian Mounted Police) in ground search and rescue operations. This assistance is usually in the form of air services support for lost and missing persons. The primary asset for this service is contracted civil aviation assets such as Universal Helicopters, Cougar Helicopters, etc. ... The CF and Canadian Coast Guard share the responsibility to respond to aircraft incidents and all marine incidents in waters under federal jurisdiction ... JRCC Halifax may provide assistance to provincial and territorial authorities in the region upon request using primary or secondary SAR assets provided the primary mission of aeronautical and marine SAR is not impacted."

JRCC staff, according to the report, decided to send two secondary SAR assets instead of one primary one, such as a Gander-based Cormorant helicopter: "In this particular case, the JRCC determined that the assets best positioned, with the right capabilities to respond to the incident in Makkovik were secondary SAR assets. Consequently, the JRCC tasked a CH146 Griffon from 444 Sqn Goose Bay and a CP140 Aurora from 405 Sqn in Greenwood (already airborne and in the vicinity of Labrador) to respond. Due to their proximity to Makkovik at the time of tasking both were able to respond faster than could either the primary SAR assets. Additionally, both aircraft were equipped to conduct night search operations ... Furthermore, by using the secondary SAR assets in this manner primary SAR assets were retained ready to respond to a second incident should it have occurred."

The report concludes with a statement about the "Appropriateness of Actions Undertaken by the CF": "The CF response was effective and contributed to the search effort, which resulted in the discovery of the teenager's body. The information indicates that commanders and staff made sound decisions in accordance with standard operating procedures using the best information that was available as events unfolded."

The report, signed by Major-General J.H. Vance, ends with an expression of sympathy, followed by a single recommendation: "On behalf of all CF members who were employed in the rescue effort, I offer our most sincere condolences to the Winters family as they deal with this tragic event. Junior Ranger Burton Winters was one of our own. Despite this heartrending outcome, throughout this incident when CF assistance was requested, the CF responded as it should have. I therefore recommend that no further investigative action is required."

The day after the report was closed, DND officials travelled to St. John's to meet with provincial politicians. Gardam met with members of Burton's family for about 40 minutes prior to his holding a press conference. "No one can imagine what it feels like to talk to a family in a grief-stricken state and say that nothing I can do, or could have done, would bring Burton back," he told reporters. "This is a tragic loss and it was one of the most difficult things I've had to do in my career."

In this press conference, Gardam reiterated why the department was late to join the search for Burton Winters, as outlined in the new report, and he discussed the details of the bad weather. At the same event, Colonel Mark Chinner, the officer in charge of the Air Coordination Component Element for DND's Atlantic region, explained that, even if the weather had been favourable, neither of the Griffons in Goose Bay were mechanically fit to fly on the morning after Burton went missing. The DND aircraft waiting in Gander and Greenwood, he added, had not been necessary because of the civilian aircraft contracted by the provincial government. "It made more sense to use the aircraft from Goose Bay at that point in time" was his succinct response.

Although the Canadian Forces properly followed all established and "time-tested" protocols during the search, Gardam added, DND might review them "in the future."

4. Changing protocols

Burton Winters's family's reaction to DND's quick and "full investigation" came quickly, and angrily. They rejected the report's findings and organized a rally to let the public know. More than 100 people joined the gathering, driving their snowmobiles onto the ice of Makkovik Harbour on the evening of February 9. Those closest to Burton voiced their thoughts. Most were concerned with the "call-back protocol" and concentrated their criticism on that apparently standard military search and rescue procedure.

"We shouldn't be standing here today," Burton's mother asserted. "They should have come out, come and helped us the first time we called. We shouldn't have had to make a second call. We didn't know we were supposed to make a second call."

Burton's stepmother, Natalie Jacque, spoke next: "We're just in utter disbelief of DND's responses and the numerous excuses that they've been giving through the past few days and it's almost unbelievable that ... [they] expect us to just let it slide, that it's first of all because of the weather, or because there was already a civilian helicopter on the scene."

"A helicopter that wasn't equipped!" Winters-Rice interjected. "They [the private volunteers] were out there helping. How come they [the military] couldn't come in?" The family was enraged that DND had left the air search up to civilians.

Jacque returned to the issue of the protocol: "It should only take one phone call for help to come ... We still feel that he's coming home to us and now that we have to deal with this on top of it all and they have no heart!"

A reporter on the scene asked the two women to state their demands. Winters-Rice wanted DND, in future cases, to respond to the first distress call they receive—not wait for follow-up pleas: "First response. When we call, you come." Jacque went further: "We want them to take responsibility for their mistake because this was obviously a mistake of theirs. They didn't come when we called and secondly we want them to have a search and rescue unit stationed in Labrador. We want it to stay here. It's needed. We don't want this to happen to any other family."

On the same topic, Burton's father, Rodney Jacque, raised the practical issue of a poor communications infrastructure:

"Isolated places need search and rescue as much or more than anyone else. We don't even have cell phone service here."

Winters-Rice had one more item on her list of demands for DND, but she didn't sound confident that she would ever receive it: "We haven't even got no apology or anything. Even an apology would be something, at least."

Elizabeth Winters-Rice, another member of the family, put her views in an open letter to the prime minister: "Dear Prime Minister Harper, I am writing a letter with regard to the ineffectiveness of the SAR handling of the detachment regarding the report of a missing youth, Burton Winters." None of the five reasons offered by DND for the long response time (communication problems, technical equipment problems, weather, local resources being used in the search, no second call) satisfied her. As she emphasized, "I feel that if anything the DND internal report has clarified many problems that have been exposed within the existing SAR procedures in Labrador. The inconsistencies and loop holes were vast enough to permit not one, but two Griffon helicopters to slip through in such an emergency. I understand that the 444 Squadron in IIVGB has SAR as their secondary role. However, what is the point in having that role if they are not available for such emergencies?"

Whereas DND considered the five reasons sufficient to close the case, Winters-Rice believed that they justified a closer look into the events and concomitant decisions. "It is my opinion that those are fine valid reasons for an independent investigation," she wrote. "The existing policies and procedures in place do not work. That proved to be evident with the death of Burton Winters and the ineffectiveness to save him. How many other people in our region will be in the same situation that will result in their demise? Please consider my plea for both an independent investigation and a fully functioning SAR in Labrador in honour of Burton Winters."

Opposition members in both the provincial and federal legislatures were quick to point out what they considered to be deficiencies in DND's explanations. They also called for the launch of an inquiry. Among those foremost in the public spotlight was Liberal MHA Randy Edmunds. His involvement in the search and his participation in the recovery of Burton's body gave him an extraordinary amount of public credibility when he spoke.

Edmunds was most frustrated by a question posed by Burton's mother: How could she have gotten to Makkovik before DND's search and rescue aircraft? But what most *bothered* him was how DND tried to answer it in its report. "The rest is history," he mused. "The rest is in the media: a non-stop request for an inquiry … Inquiries have been called for less serious issues and there are still issues of protocol that have to be addressed."

Amidst all the accusations and letter-writing, DND quietly reviewed the protocols in question and its officials had generated an internal recommendation about changing them. This recommendation, submitted to DND on February 17 in a letter by Lieutenant-General Walter Semianiw, Commander of Canada Command, was publicly released on March 8: "In response to your request for a review of current protocols related to CF support to Ground SAR (GSAR) events, which are the responsibility of law enforcement agencies, Canada Command conducted a series of consultations regarding the interaction of our Joint Rescue Coordination Centres (JRCC) and Provincial and Territorial authorities responsible for missing person searches, as defined in the Reference."

Semianiw had discovered little wrong with the existing

system. The second-call protocol was official and widely observed, but did not exist in writing. His assessment: "It is clear from the review that the existing protocols work well, with clear direction and procedures specified and tested over time ... Of the many issues considered the only area where we may further strengthen an already strong system, relates to identifying the most effective times for communication between CF and GSAR authorities for a GSAR event; more specifically, identifying the process whereby the CF closes a GSAR case file, and how this is performed and communicated to the requesting agencies and/or Other Government Departments (OGDs). Historically, this has not been a problem even though the protocol has not been in written format."

With no written protocol to amend, Semianiw recommended changes to the unwritten protocol: "If unable to assist [following an initial plea for help], JRCC will explain why this is the case and ask the RA [requesting authority] to re-establish contact (normally a call back) at a suitable time based on the conditions and circumstances affecting this situation. The case file will remain open throughout this period. As operational conditions permit, and on a periodic basis should the case remain open for a lengthy period of time, the JRCC should contact the RA to receive an update on the situation and review the needs assessment if the situation warrants such a review ... Prior to JRCC closing the case, a confirmation call with the RA will be made to ensure that no further assistance is required. The JRCC case file can not be closed until this positive-action 'handshake' is completed between JRCC and the RA."

On March 8, almost three weeks after Semianiw's recommendation was presented to government, MacKay and Penashue held a conference call with media to advise them that a new protocol was forthcoming. However, MacKay implied, the country's search and rescue system worked perfectly well and, if there was any fault to be found, it wasn't with the federal government.

"Following the tragic death of young Burton Winters, I suggested and requested that the Canadian Forces conduct a review of the military protocol with regards to ground search and rescue," MacKay explained. "This review and consultation with our partner agencies, including the Fire and Emergency Services of Newfoundland and Labrador, the Royal Canadian Mounted Police, the Royal Newfoundland Constabulary and the Newfoundland Department of Justice is now complete. The result of this review is an amended protocol whereby the military will provide an additional layer of due diligence. That is to say that in the search for Mr. Winters and prior to this review the onus was on provincial authorities, the legal authority and lead for ground search and rescue, to call the military and request additional assets if required ..."

Following the military review that MacKay requested, "military officials will now proactively call back the lead agencies to ascertain whether Canadian Forces assets are still requested. In Newfoundland and Labrador this protocol is in place and we intend to extend this protocol nationally. I believe the adoption of this new protocol improves communications and enhances situational awareness among all ground search and rescue agencies and working together we will continue to ensure one of the best search and rescue systems in the world."

After the press conference, MacKay officially announced the protocol change in the House of Commons. He prefaced his remarks with a disclaimer about Ottawa's responsibility in the search for the Makkovik youth.

"While the legal authority for ground search and rescue does rest with the provincial and territorial governments, the Canadian Forces, as a partner in the search and rescue network, nevertheless, reviewed the protocol surrounding Canadian Forces participation in this search," MacKay stated. "In future, the Canadian Forces will implement a call back protocol to ensure continuous communications on ground searches in order to enhance awareness of changing circumstances and the potential need of Canadian Forces participation. This ongoing dialogue will continue until the file is closed. This new protocol will enhance the capabilities of all partners to our search and rescue network across Canada."

DND officials must have been sorely disappointed in the public's reaction to this protocol change. Although this change had been one of the main demands made by Burton Winters's family, they were not as grateful as DND might have wished.

"I'm not very pleased with receiving just 'a second phone call won't be needed,'" Natalie Jacque stated publicly. "We're glad that there's something. This possibly will be able to help others in the future, but I mean, it's something so simple it should have been in place before anyway."

Other critics spoke up immediately, including one of the province's leading experts on search and rescue: Clarence Peddle, a retired SAR coordinator, who made his opinions widely known on the Burton Winters Rescue Centre IN Labrador Facebook page. He posted a statement on April 3, 2012, that revealed little faith in any of the explanations by the Canadian Forces or in the value of their protocol change. His post contained a series of questions: "Is it possible that the failure to deploy aircraft was really the result of a SAR operation that was responded to in a manner that deviated significantly from the due diligence that it deserved? Is it possible that the response fell short because the people involved in the decision-making process did not have a competent understanding of their role and responsibility? Subsequent to this incident, SAR officials came out with a new protocol on callback. Is it really a new policy or just an acknowledgement of what should have been done initially: is the 'new protocol' merely a pretext designed to mask this failure?"

In 2006, Peddle participated in a meeting involving federal and provincial agencies to discuss and repair the problems experienced during "another multi-jurisdictional operation." During that meeting, Peddle described in his post, the agencies clarified their roles and jurisdictions and "strengthen[ed] the communications so that nothing fell through the cracks." They did not, however, establish any kind of protocol, written or otherwise, that would require a requesting authority to make a second call if the first was refused. "I will say categorically that at no point did anybody mention the call back protocol as suggested by DND to be in effect for the search for Burton Winters. I retired in 2009. I did not see or hear any change in SAR's policy regarding humanitarian incidents following the 2006 meeting up to that point. Could this 'time-tested policy' have come in effect in the last three years, then?"

Considering the extensive volumes of JRCC written protocols, Peddle wondered why only the "call-back" protocol remained unwritten, with no record of when, how, or by whom it was officially adopted as policy. "But the main reason I would say that no such 'call-back policy' existed is because I cannot imagine anyone working in Search and Rescue implementing such a hard and fast, dangerous and senseless protocol in the first place, a protocol that would drive a stake through the heart of the National Search and Rescue plan," he concluded. "It would have the effect of providing a lower level of service within a SAR system, which could in some ways already be described as second class."

Barry Andersen, Makkovik's SAR coordinator, was the one man, besides Cpl. Vardy, who should have known if he had to make a second call for help. But Andersen had never heard of that protocol either: "No, I haven't—not as a ground search and rescue coordinator here." Asked if he thought such a call-back requirement would make sense for search and rescue, Andersen responded, "Not to me, no."

No officials from either DND or the Canadian Forces have publicly addressed the criticisms levelled by Andersen or Peddle. No further details about how the protocol came to be implemented or why it was never written down have been offered.

5. How the Marine Rescue Sub-Centre fits in

The Canadian Coast Guard's Marine Rescue Sub-Centre in St. John's played no part in the search for Burton Winters, but it did play a role in the public debates that erupted following the discovery of his body.

The sub-centre was set up at the St. John's Coast Guard base in 1977, one of two in eastern Canada designed to assist the JRCCs in Halifax, Nova Scotia, and Trenton, Ontario, by providing local knowledge and expertise and sharing the workload. They were part of a larger system described by the Coast Guard in a Notice to Mariners on Search and Rescue in Canadian and Adjacent Waters (prior to June 2012 amendments): "The Canadian Forces (CF) in cooperation with the

Canadian Coast Guard (CCG) has overall responsibility for coordination of federal aeronautical and maritime Search and Rescue (SAR) activities in Canada, including Canadian waters and the high seas off the coasts of Canada. The CF provides dedicated SAR aircraft in support to marine SAR incidents. The CCG coordinates maritime SAR activities within this area and provides dedicated maritime vessels in strategic locations. Joint Rescue Coordination Centres (JRCC) are maintained at Victoria, B.C., Trenton, Ont. and Halifax, N.S. These centres are staffed 24 hours a day by Canadian Forces and Canadian Coast Guard personnel. Each JRCC is responsible for an internationally agreed designated area known as a Search and Rescue Region (SSR) ... In addition, Maritime Rescue Sub-Centres (MRSC), staffed by Coast Guard personnel are maintained at St. John's, Nfld. and at Quebec, Que., to coordinate local maritime SAR operations."

JRCC Halifax was established under the Royal Canadian Air Force in 1947 as one of two Rescue Coordination Centres—the other was in Torbay, Newfoundland. At the time they worked with military air rescue units stationed in Torbay and in Greenwood and Dartmouth in Nova Scotia. Together they were responsible for the Atlantic SAR area, one of four areas in Canada. In the first year of operation, they dealt with 31 of the 50 total search and rescue incidents that had taken place in the whole country.

Several decades and administrative reorganizations later, the Halifax Search and Rescue Region (SRR) is still the busiest zone in Canada for maritime emergencies. JRCC Halifax is responsible for approximately 4.7 million square kilometres of ocean and coastline territory stretching from the southern half of Baffin Island to the fishing banks south of Nova Scotia and from mid-Quebec to halfway across the Atlantic

Ocean.

"Joint Rescue Coordination Centre (JRCC) Halifax is the focal point of all aeronautical and maritime SAR activity within its region," the JRCC Halifax website explains. "The staff collects and distributes essential information concerning a distress situation, arranges the dispatch of rescue assets and personnel to ships or aircraft in distress and coordinates the efforts of all responding resources."

At the time of its closure, the Marine Rescue Sub-Centre in St. John's had 12 employees, two of whom were Marine Coordinators directly involved in search and rescue. Their jurisdiction covered the entire length of the Labrador coast and all around the island of Newfoundland: 28,956 kilometres of coastline and more than 900,000 square kilometres of ocean. In June 2011, DFO, which shares responsibility for the Coast Guard, reported that the St. John's sub-centre responded to an average of 421 calls each year—25 per cent of which were made by people in serious distress. Although JRCC Halifax dealt with 1,912 incidents annually, only 10 per cent were distress cases.

Jason Hamilton, a former employee of JRCC Halifax, said those numbers have been going up steadily, from 2,212 incidents within Halifax's jurisdiction in 1994 to 2,868 in 2010. Statistics published by DND show that the numbers fluctuated widely in that period, from a low of 1,914 calls in 2004 to the high recorded six years later, but they, too, indicate a general increase in maritime and aeronautical distress calls off the east coast of Canada. Twice (in 1999 and 2007) the number of cases had risen close to or topped the 2,800 mark.

Nevertheless, in 2011 the federal government cut more than $50 million from DFO's annual budget, propelling the Coast Guard to close both the St. John's and Quebec City sub-centres and to consolidate the services of the Newfoundland facility in Halifax with the JRCC. The Canadian Forces maintained the action was justified, given modern improvements to communications.

Conservative MP Randy Kamp, then Parliamentary Secretary to the Minister of Fisheries and Oceans, said in the House of Commons on June 20, 2011, that the cuts would eliminate an inefficient "duplication of services" for the benefit of the Coast Guard. "The main point I want to make is that the decision to consolidate was a careful decision made on the recommendation of the Canadian Coast Guard and it will not compromise the on-water response time," Kamp stated. "By transferring resources to the joint centres, it will make it easier for the Coast Guard to work more closely with its Canadian Forces partners by locating all maritime and air search coordinators in the same centres ... It goes without saying that the rescue centres will continue to be operational 24 hours a day, seven days a week, and staffed by Canadian Forces and Canadian Coast Guard personnel who are thoroughly trained to evaluate various situations and send the most effective resources to deal with a particular incident. We will continue to ensure that local knowledge and expertise are embedded in the tools and training of the crews, mariners and Coast Guard employees."

The media found out about the planned cuts before the official announcement; by the time the decision was confirmed by Parliament, criticisms were already being voiced. The critics, many of whom had extensive search and rescue experience, argued that the scaled-back system would be less efficient and risk people's lives.

Maurice Adams, a retired Coast Guard watch supervisor from Paradise, Newfoundland, spoke out strongly against the cut. He charged that the Canadian Coast Guard was choosing financial efficiency over operational efficiency. "The St. John's centre is called the Maritime Rescue Sub-Centre for a reason," Adams stated in a letter to the *Telegram* in August 2011: "While air/marine coordination efficiency is important (and which is what is now being by senior bureaucratic and political officials as something whose efficiency will be improved), its improvement is largely a red herring. Notwithstanding that the maritime coordination, expertise and efficiency of the St. John's Rescue Sub-Centre is key, it is second to none and cannot be improved upon (only degraded) by taking it out of the Coast Guard's Regional Operations Centre and moving it to Halifax."

The St. John's sub-centre, Adams continued, placed all the individuals who were needed to coordinate a marine search and rescue in the Newfoundland and Labrador region in one room. There was "[n]o middle men/women, no separation by a 1,000 miles of ocean, no language barriers, no loss of vessel movement, ice operations or safety oriented situational awareness. It is the coordination of search and rescue from this 'maritime' perspective and it is in this team-oriented, horizontal organizational structure that makes the St. John's Maritime Rescue Sub-Centre unique and not only the most efficient, but the most effective."

Hamilton expressed his concerns directly to MacKay in a letter in early 2012: "Although the Coast Guard does not fall under DND, it must answer to the Department with respect to its role in search and rescue. As the Minister in charge of National Defence, I hold you personally responsible for the repercussions if this amalgamation is allowed to continue." He was particularly concerned that the planned elimination of seven to 10 east-coast SAR specialists, as well as of the

Marine Rescue Sub-Centres, would make the provision of search and rescue services unmanageable: "I consider these cuts irresponsible and dangerous. The employees of all three centres [JRCC Halifax and the sub-centres in St. John's and Quebec City] are unanimously in agreement with these concerns."

The closure of the St. John's sub-centre, Hamilton pointed out, would mean that its workload would fall to JRCC Halifax, which was facing staff cuts and would be unable to effectively meet the new, additional demands. At the busiest times the case load would likely overwhelm the Halifax centre; additional relief staff would need at least half an hour to get to their posts, an unacceptable length of time in an emergency. "The addition of 30 minutes to the current response times of our resources will be compounded and likely lead to unnecessary loss of life," Hamilton asserted. "The public statements made by DFO that 'technology has changed significantly and will enable these cuts to occur with minimal impact' again either demonstrates ignorance or a covert attempt to downplay the effects of these cuts."

St. John's sub-centre employee Merv Wiseman worried that the amalgamation would increase response time and waste precious minutes. His summation: "Seconds count. If you're going to apply the rationale that it's more convenient there's efficiencies in consolidation and so on, I have to ask the question: 'Why are you stopping at Halifax?' If we're going to use that rationale, maybe we should move it into Winnipeg, bring in the JRCC in Victoria, bring in Trenton and Halifax and put them all into one centre."

Local politicians had their say. Dennis O'Keefe, mayor of St. John's, bluntly called the closure of his city's rescue sub-centre "stupid." He announced that he would write Prime Minister Harper with his thoughts and he invited everyone else in the province to follow his example. The provincial PC government made similar pronouncements; Fisheries Minister Clyde Jackman urged citizens to rally and to express their disappointment. "From my perspective it's a life safety issue," Jackman said. "When you're looking at the fiscal decisions you're making, surely there's other places where you can take people out, rather than removing people [where] such a decision could impact the safety of people."

In Ottawa, the federal Opposition parties repeatedly asked the government to reverse its policy—to no avail. "At this point in time, my concern is that we cannot seem to get across to the Government of Canada how important that sub-centre is to the lives of people who spend much of their time on sea, whether we are talking about sailors, fishers, or people who just use the sea for pleasure, or oil workers, for instance," Liberal MP Judy Foote declared in the House of Commons on June 20, 2011. "We have been trying to tell the minister responsible for Fisheries and Oceans and the Prime Minister and anyone else who will listen that to continue down the path of closing that sub-centre is going to mean much more harm to people. We have said time and again that the people working in the sub-centre really need to know Newfoundland and Labrador."

NDP MP Jack Harris rebuked federal officials for implying that the facility was less vital than it actually was: "I think it's a shameful decision. The minister [Keith Ashfield] continuously referred to the marine rescue coordination centre as a call centre, downplaying the significance and importance of its role."

Within days of the announcement of the closure of the St. John's sub-centre in 2011, 2,000 people protested on the St. John's waterfront, demonstrating that it wasn't just a few politicians and disgruntled ex-employees who thought closing the sub-centre would be disastrous to the province. Waving signs that screamed "Safety first," "Save our rescue centre! Save lives!," and "Shame on you, Harper!," the protestors had eagerly responded to calls from a committee of activists, union leaders, and politicians to organize. "I'm pretty angry, coming from a long line of fisher folk," said protester Ellen MacPherson, whose brothers had followed their father to make a living from the sea. "It just breaks my heart to see that they're taking away something that's so vital."

Despite the widespread and persistent outcry, Ottawa refused to change a single detail of its plan to close the sub-centres until the week Burton Winters was lost. Whether by cruel coincidence or cold calculation, the government chose February 1 to announce it would be closing the St. John's sub-centre two months earlier than scheduled—in April instead of June. The news did not become widely known for almost a week, but, when it did, the reception was negative. One provincial editorial labelled it "Adding insult to injury": "Just when you thought the Conservative government couldn't get any more ignorant in its decision to 'streamline' the country's search and rescue operations, they once again outdo themselves," thundered the weekly *Southern Gazette* on February 6, 2012. "The news of the acceleration was made on the very same day military search and rescue aircraft found the body of a 14-year-old boy from the northern Labrador community of Makkovik."

That was, in fact, the only connection between the St. John's sub-centre and the search for Burton, but people found more. One of the first to do so was a Canadian Forces officer who had made decisions during the search for Burton.

In a telephone conversation recorded on the morning

of February 2, punctuated by laughter and remarks such as "Yeah, we hear you!" and "Right on!," Major Ali Laaouan, JRCC Halifax's OIC, called the sub-centre and spoke with a man named John on the Air Coordination desk. Laaouan told John to update a "media tape" with the details of the recently completed Makkovik operation, saying that he should make it plain that the search for Burton Winters was outside of the JRCC's jurisdiction, that it was entirely the responsibility of the provincial emergency measures organization.

"Basically that we, you know, that we helped out. You know: we were called on this day at this time and we provided the helicopter from Triple-4 Squadron and an Aurora that has the, I guess, the proper ..."

"FLIR," John interrupted—the acronym for Forward Looking Infrared Radiometer.

Laaouan continued: "... things for, yeah, the FLIR and all that stuff ..." He made it clear that he would not be doing any interviews with journalists and he didn't want anyone else at the sub-centre to, either: "I know what they're trying to do and I don't want to get into that.... There's a lot of hype now with the media about a bunch of stuff. I'm not going to entertain any requests for media or anything like that because they're going to try to tie it [the search for Burton Winters] into the sub-centre closure and all that."

Laaouan was correct, but it wasn't just the media that was scrutinizing the JRCC. The unions representing the SAR employees—principally the Public Service of Canada (PSAC) and the Union of Canadian Transportation Employees (UCTE)—were at the forefront of criticism from the time the cuts were first revealed. The unions let employees like Wiseman speak out without fear. "Without the protection of the union, my voice would have been silenced on this issue long ago," Wiseman clarified in August 2011. "In the department we are not permitted to speak publicly without employer approval, but we at PSAC-UCTE are taking this fight on for the greater good."

In February 2012 the unions had to respond to the expedited closure of the St. John's sub-centre. "I don't understand why the department is determined to put Canadian lives at risk," challenged Christine Collins, UCTE national president. "Since the first announcement we have asked for an independent impartial review of the announced closures. If the department is so confident that lives will not be put at risk, then they should not fear an independent expert review."

Even as the federal government was closing the St. John's and Quebec City sub-centres, the Canadian Coast Guard was applying to have the search and rescue coordinators in both centres declared essential service workers under the federal Public Service Labour Relations Act. The Coast Guard wanted them to be "necessary for the safety or security of the public," in order to prevent labour disruptions.

"It's like they don't know what they're doing," Collins said. "On the one hand the union is being told that should there ever be a strike these services are essential—and we agree. But on the other hand, CCG is saying that these services are not so essential for the safety of Canadians in order to close them."

Meanwhile, the Coast Guard revealed that they were having trouble finding properly qualified candidates to fill positions at Halifax and Trenton JRCCs, so they were more flexible about qualifications. Instead of requiring the standard Watchkeeping Mate Certificate, it would accept the less stringent Maritime Naval Officer Certificate for its employees.

"What are they thinking?" Collins asked. "In no way is a Naval Officer Certificate equivalent to the Transport Canada Watchkeeping Mate Certificate. SAR coordinators are put through a series of rigorous testing that demonstrates an ability that is unparalleled. Naval Officer Certificate is an inappropriate equivalent. It's like saying just because I go kayaking, I'm a qualified mariner."

On March 21, 2012, the unions took the protest from the St. John's waterfront and brought it to downtown Ottawa. PSAC organized a "reception" on Parliament Hill so that select federal public servants could offer advice to the Conservative government about the March 29 budget. Other issues, such as food safety and the needs of veterans, got a thorough airing, but not before Wiseman again entreated the government not to close the St. John's and Quebec City rescue sub-centres: "It's about public safety."

Wiseman did not specifically mention the search for Burton Winters, but he had brought along a survivor from a SAR operation near Cape Harrison—a man who couldn't fail to remind the audience about the other Labradorian who had not been so lucky. "Nobody knows when you are going to run into trouble," said Todd Broomfield, one of three Makkovik fishermen who had been forced to abandon their boat just before it sank in rough waters. They had taken refuge on shore by crawling out of the ocean and scaling partway up a tall, steep cliff, where they had perched for almost six hours. One of the men had used a satellite phone to send a distress call to his wife and DND had sent a Cormorant from Gander to find them and pluck them off the rocks to safety.

Broomfield told the gathered civil servants that the impending closure of the St. John's sub-centre would adversely affect people living in remote parts of his province: "Newfoundland has a long coastline. It's exposed. It's rugged. Surely our government can afford to keep a marine rescue centre in

St. John's that provides such a valuable service to the people who make a living in this harsh environment. Their knowledge of Newfoundland and Labrador is second to none. We don't want to lose that ... This knowledge is priceless."

PSAC demanded that the planned changes to the country's search and rescue infrastructure be put on hold until a review of their possible social and economic impacts could be conducted. The government responded by attacking the union. "We are not surprised that self-serving union bosses would once again tout a plan to raise taxes, hike spending and increase the size of government," Sean Osmer, a spokesman for then Treasury Board president Tony Clement, wrote in an email to reporters: "Our government has remained focused on a low-tax plan that reflects the priorities of Canadians—jobs and economic growth. Our responsible plan makes government leaner and more affordable; it will boost the economic recovery and will be good for job creation."

In spite of vigorous lobbying, the protests in reaction to the Burton Winters tragedy and the cuts to search and rescue seemed about to stall in the face of an intransigent federal government. Federal officials may have thought they had adequately dealt with or deflected the issues. They may have imagined they'd reached a point when no further action was required, but that would soon change, in a large part because of the CBC.

6. *The Fifth Estate* has a say

On March 23, 2012, two days after the PSAC reception in Ottawa and seven weeks after the search for Burton Winters ended, *The Fifth Estate*, a CBC current affairs program, aired a segment called "Lost on the Ice"—an hour-long examination of the official response to the Burton Winters emergency. Its blunt and specific criticisms of DND's actions and decisions added fuel to the languishing debate, bolstering those who demanded that someone be accountable for the tragedy and that improvements to search and rescue be made. "A call for help unanswered," was the subtitle of the episode, which concentrated on a single question: "Why did the Canadian Forces refuse to send the search and rescue helicopter that could have saved a teenaged boy lost in a Labrador blizzard?"

Using interviews with family and others involved in the search, showing family video footage of young Burton, and staging re-enactments that participants said were chillingly accurate, *The Fifth Estate* told Burton's story. The program focused public attention on the accusation that the military may have been able to save the youth's life—but instead made decisions that removed all chance of his being found in time.

"Lost on the Ice" began with a description of Burton and his ordeal. His father and stepmother are introduced, as is his little brother, his dog, Quinn, and other members of the community. The program documented how the search team reacted to the unwelcome news it kept getting from the JRCC about unavailable military air support, and it pointed to a series of inconsistencies between what DND said had happened and what seems to actually have taken place. It raised questions about the weather, about protocol, about federal/provincial jurisdiction, about holding the Gander Cormorants in reserve for another emergency, and about the Canadian Forces' numerous broken-down aircraft.

Reporters from *The Fifth Estate* tried to get information from the proper officials, those who had already spoken about the issues surrounding the search, but had no success.

All they got back from their inquiries was a short email from a naval public relations officer. "We repeatedly asked Defence Minister Peter MacKay and Rear Admiral Dave Gardam, the commander of First Task Force Atlantic, to explain these inconsistencies in an on-camera interview," CBC reported. "Both declined."

Public anticipation for *The Fifth Estate* episode ran high in Newfoundland and Labrador and among the 36,000-strong Burton Winters' Rescue Centre IN Labrador Facebook group. In the days and hours before it aired on March 23, many members were compelled to alert one another that the program was coming up and to announce their intentions to watch it. A roll call of sorts took place.

One member wrote that she was "gonna light a candle while we watch *The Fifth Estate* at nine!" Another: "It's gonna be rough watching the re-enactment for viewers, but I can only imagine what his family is going through!" The messages continued as the episode was under way: "The show wasn't on five minutes and my tears were flowing. He showed such strength, determination. Long may your wonderful spirit live on!" Other posts included: "Watching *The Fifth Estate*. I may not have known Burton, but his story brings me to tears. Rest peacefully Burton." "Been waiting all day for *The Fifth Estate* and now it's on I'm heartbroken. This story from the beginning has been so sad. RIP Burton." "I'm absolutely heartbroken watching this. May he rest in peace. Heaven sure did receive an angel."

After *The Fifth Estate* episode aired and its emotional impact had settled, the online remarks displayed a measure of pointed anger. "Sorry Peter MacKay," one viewer from Happy Valley-Goose Bay had written as the final credits rolled. "If your child was being bullied and got killed in the playground because the teacher was in the front of the school 'just in case' someone might fall in the driveway, would you think it was okay? 'Just in case' is not good enough." "They have blood on their hands," was another comment. "This province continues to be ignored when it come to adequate SAR. Enough already! RIP Burton." And another: "An utter national shame. I could have at least had some respect for our government if they were willing to acknowledge their shortfalls, mistakes, ridiculous protocol, and poor decision-making—at least have the face to own it! I don't point blame at any one person, but [at] our Canadian government as a whole. You failed Canada. So sad thinking about Burton and his family tonight."

One of the most poignant messages was left by Elizabeth Winters-Rice: "I've stated from the beginning that the DND report had inconsistencies. It has now been proven true. So happy that it's now official, but sad that we are supposed to be in a democratic society and our government denied and lied all the way through!"

For hours and days after the program had aired, the Burton Winters Facebook page filled with messages echoing the same sentiments.

"Lost on the Ice" was mentioned during the next sittings of both the provincial House of Assembly and the federal House of Commons. In St. John's, Liberal Dwight Ball, acting leader of the Official Opposition, had a pointed question for the government: "*The Fifth Estate* documentary on Burton Winters has rocked the province by revealing some very shocking information around protocol and DND guidelines. It is unfortunate that it took *The Fifth Estate* to uncover those disturbing facts. My question is to the Premier: 'Did your investigation following this incident identify any of those shocking revelations?'" Premier Dunderdale answered in the negative: "Mr. Speaker, our investigation was around our own protocols and a review of those protocols to see if we need to be doing something else provincially. Mr. Speaker, I, like others, accepted that weather was an issue. *The Fifth Estate* has made it quite clear that weather was not an issue. There is a letter leaving the minister's office today to the Minister of Defence asking the question: If weather was not an issue, why weren't the Cormorants deployed? The answer to that question, Mr. Speaker, will dictate further actions by this government."

Ball, Edmunds, and Cartwright-L'Anse au Clair MHA Yvonne Jones continued to press the premier, using *The Fifth Estate* revelations to ask again for a provincial inquiry. "I watched, along with the people of Makkovik, in disbelief and shock Friday night as *The Fifth Estate* reported the true facts and shortcomings of DND's mishandling of the terrible tragedy," Edmunds said. "If DND is not prepared to do a full investigation, my question is to the Premier: Will your government step up, do what is right, and get the real information for the families and friends of Burton Winters?"

Dunderdale fully understood the outrage at the explanations DND had offered, but, she added, she could do little beyond asking questions of the federal government.

Jones, then Liberal House Leader, asked Dunderdale if she would "instigate a judicial inquiry which would allow the access to information that we need to know to see what happened in this particular incident on search and rescue?"

Dunderdale repeated her promise: "Mr. Speaker, the House Leader well recognizes that I have no authority to institute an inquiry into the federal government's activities to have access to the kind of information that we would need. I can call on the federal government for such an inquiry. That may very well happen, Mr. Speaker. First of all, I am demanding an explanation from the Minister of Defence with regard to the information that was in *The Fifth Estate* program. Once I receive the response to that inquiry, Mr. Speaker, I will make it available to the people of the province."

Federal officials may have been able to dodge *The Fifth Estate* before the episode aired, but afterwards, they could no longer completely avoid the questions it raised. MP Jack Harris was the first to bring up the CBC program in the House of Commons: "Mr. Speaker, CBC's *The Fifth Estate* has uncovered disturbing facts about the Canadian Forces' response to the search for Burton Winters in Makkovik in January. His family described the military's explanation as, 'One excuse wasn't enough for them, they had to give five.' It was not the weather, it was not the protocol. They closed the file and later said they had no equipment available to do the search. One former SAR coordinator called the CF report 'abysmal, misleading and wrong.' What is the state of our search and rescue system in Canada? Will the government establish an independent inquiry to find out the full truth about what happened and what needs to be done to protect Canadians like this boy in Makkovik?"

Although MacKay rose to respond, he did not directly address Harris's questions. Instead, he recapped the federal government's version of events and reminded the House about what his department had already done: "It was certainly a tragedy," he began, "this young man was a member of the Canadian Forces junior rangers program. Members of his troop assisted in his search. As the member [Harris] would know, the reality is that the first call to the Canadian Forces came some 20 hours after young Mr. Winters was last seen. The second call was placed 51 hours later and the Canadian Forces assets were deployed. We have improved the protocol with respect to the communications between the province and the federal government and that protocol has ground search and rescue responsibility with the province."

Liberal MP Judy Foote of Random-Burin-St. George's used CBC's revelations to link the search for Burton Winters to Ottawa's plans for closing the sub-centre in St. John's: "The real reason a search and rescue helicopter was not sent to Labrador to find missing 14-year-old Burton Winters is now clear. The rear admiral said he could not spare a helicopter in case it was needed elsewhere. This tragedy should never have happened. In spite of this and the lack of search and rescue resources when needed, the government is closing the marine rescue sub-centre in St. John's with its local knowledge and expertise. Because the member for Labrador [Peter Penashue] will not, will someone in the government tell the Prime Minister he has to reverse his decision or more lives will be lost?"

No one did. Fisheries Minister Ashfield spoke next: "Mr. Speaker, as I have said many times in the House, we would never close the sub-centre in St. John's if we thought we would be putting mariners or anyone else at risk." Therefore, "[w]e will continue with the process of closing the sub-centre."

Foote and Harris presented separate petitions to the House on behalf of those who wanted the sub-centre to remain open. Foote's petition held the names of 1,300 Newfoundlanders and Labradorians; Harris's had been signed by citizens from across the country. Neither petition changed the mind of the Conservative politicians.

That doesn't mean *The Fifth Estate* episode did not have

an impact on the federal government. Shortly after it aired, Penashue broke a lengthy silence to address CBC's coverage and conclusions, albeit cryptically. He did not respond directly to the charges, but told reporters that the CBC had not conducted proper research before releasing the story: "There is an angle that *The Fifth Estate* took and unfortunately the facts weren't sought, which is very, very, I guess, unfortunate ... because it doesn't present a fair, full picture of exactly what happened. My view is that you need to ask the right person that question ... because if you ask someone that question—to the wrong person—you're going to get a response that's not current."

At no point did Penashue identify the "right person." Nevertheless, the Labrador MP's answers seem to be the only ones ever offered by the federal government on what *The Fifth Estate* exposed to the general public.

7. To fly or not to fly

The Fifth Estate's "Lost on the Ice" episode did not result in any decisive action from the federal government, but it amplified the call for an independent public inquiry into the search for Burton Winters. It also clarified many of the issues that such an inquiry could be tasked to investigate.

Questions still lingered over the call-back protocol—such as whether it ever existed—but other aspects of the incident came to the forefront and demanded more attention. The main question on the public's mind has been asked repeatedly since January 30, 2012: Regardless of all other considerations, why didn't the Canadian Forces send air support, a primary SAR asset, on the Monday morning when it would have done the most good and might even have saved Burton's life?

Retired SAR coordinator Peddle was definite: "I do 100 per cent believe that Burton was indeed alive well into Monday afternoon. Furthermore, I am 100 per cent certain that if a cursory helicopter search had been conducted Monday afternoon, then they would certainly have found the snowmobile and Burton."

The two reasons given for not sending the aircraft—bad weather and faulty equipment—had been heard many times but never fully accepted by Burton's family and their supporters. They were certainly not accepted by Peddle, who accused the military of hamstringing itself with a harmful fixation on procedural details. The JRCC should have sent an aircraft right away, he maintained, even if the weather conditions had been worse: "Nobody would legitimately criticize Halifax for sending a helicopter to save a kid that was lost in a blizzard."

That wasn't DND's approach. In his press conference, Rear Admiral Gardam had said that, "[g]iven the weather conditions, which were below limits for safe operations of an aircraft, our aircraft were not able to operate in that environment." A few weeks later, in early March, MacKay had expressed the same sentiment: The weather had been the main reason support was not sent immediately, and, when it cleared, nothing stopped the Canadian Forces from getting involved—once they received a second call for help, of course. "In the search for Mr. Winters I'm informed the weather conditions in Makkovik did not meet the minimum conditions. When the weather improved the Fire and Emergency Services of Newfoundland and Labrador requested civilian assets which were then able to respond approximately three hours later. Fire and Emergency Services of Newfoundland and Labrador did not call the Joint Rescue Coordination Centre again until 4:54 on the 31st of January."

However, the initial report given to JRCC Halifax on the Monday morning of the search that "weather in the area will not permit launch of local helo or aircraft" was followed a short time later with a report that Makkovik and vicinity had a 600-foot cloud ceiling and visibility for a full mile—twice the DND guidelines for minimum weather requirements for SAR helicopters to take to the air.

Later in March 2012, Gardam's public relations officer emailed CBC in an attempt to explain the discrepancies: "The weather conditions at Makkovik (120 nm NE of Goose Bay) were marginal at best at the time of the call with a ceiling of 600 feet and visibility of 1 nm, with overcast conditions and light snow. In fact it was the weather that initially prevented the FES-NL contracted helicopters from participating, prompting FES-NL to contact JRCC. Due to the requirement for the Cormorant to maintain the Aeronautical and Maritime SAR response in the Newfoundland region, the CH146 Griffon from 444 Squadron Griffon is a secondary SAR asset for JRCC and, in accordance with the 1 Canadian Air Division orders, secondary SAR assets are subject to different weather minimums than the Cormorant ... These limitations are 500 feet above any water or ground, and visibility of one nautical mile and clear of clouds. The reported actual conditions from Makkovik at the time were: Wind of four knots from the south-east, visibility of one statute mile in light snow and with an overcast ceiling at 600 feet."

Glossing over the fact that both Goose Bay Griffons were out of service, Gardam's officer explained that the weather had prevented both of them from flying: "The CH146 Griffon helicopter from 444 Squadron in Goose Bay is especially susceptible to weather conditions which could cause icing on the rotor blades since it has no de-icing capability. If this aircraft was caught in icing conditions it could potentially lose the ability to fly. This factor was also a consideration."

The JRCC incident log, however, tells a different story, suggesting that the weather reports had played little role in the OIC's decision not to send a Cormorant helicopter from Gander, which was the only military aircraft available to help and was, in any event, probably the best aircraft for the job. "Discussed with OIC and based on no serv Herc he would only be willing to commit 444 Squadron," C. Macdonald recorded at 9:18 a.m. on January 30, indicating that he would not send a Cormorant because none of the fixed-wing Hercules stationed in eastern Canada was fit to fly or could provide the helicopter with extra support. Minutes later, after telling Laaouan that visibility was 1 mile with a 600-foot ceiling (within the Cormorant's guidelines), Macdonald noted: "At this point he does not want to commit resources other than 444 and they are US."

Three minutes later, Macdonald was on the phone with FES-NL, but he said little about unserviceable aircraft and nothing about the need to reserve aircraft for a possible emergency in Newfoundland waters. Instead, he "advised based on weather and aircraft status [they] cannot support at this time."

According to the log, Laaouan had first decided to withhold the Cormorants based on the fact no serviceable Hercules was available. His only knowledge of the weather conditions came from a vague report from Makkovik that indicated a private aircraft couldn't take off. His reason for not being "willing" to send anything but a 444 Griffon is not explicitly recorded. When Laaouan received a more favourable weather report 10 minutes later, he also learned that neither of the Griffons could fly, but he did not change his mind about the Cormorants. There was no further recorded discussion about sending one of the Gander helicopters until later the next day. Again, Laaouan stuck to his guns.

"He wants 444 to go if they are serv in next hour or so and if not send Aurora," Macdonald wrote in the log. "He does not want to send the Corm with no serv Herc in the region." No mention of weather.

Why did Laaouan make this decision, which does not seem to have been sanctioned by any written protocol? Cormorants usually fly on search and rescue missions with the fixed-wing Hercules as backup, providing higher altitude surveillance and a different suite of sensing equipment. Perhaps Laaouan truly believed the inclement weather posed too great a risk to the aircraft and its crew without the Hercules in attendance, but, if so, it isn't reflected in the official record. A closer examination of that one decision by Laaouan could reveal more information about the process and the rationale behind it, but Laaouan had already declared he would not speak with news media.

"They couldn't send it because somebody else might call!" is how one emotional elder expressed his bewilderment during a public meeting in Makkovik. "You know what that says to us? The life of a 14-year-old Inuit boy is second to somebody else who may call. That's discrimination. That's disgraceful!"

8. The RCMP under fire

The RCMP's efforts and conduct during the search for Burton Winters were, for the most part, laudatory. The leadership of Cpl. Vardy in coordinating the dozens of volunteers who combed the wilderness around Makkovik for the lost boy has garnered nothing but praise, from officials in both levels of government and from the people of Makkovik. As well, no one has forgotten that it was a fixed-wing RCMP aircraft that made it possible for the searchers to find Burton's snowmobile—an aircraft that had not even been detailed for the task but had flown into Makkovik simply to deliver divers and underwater search equipment.

The federal police force, however, has not escaped all criticism and questions. The complaint against the RCMP came in early May 2012: It was accused of not trying to arrange for a search aircraft until the Monday morning of the search. "The fact that they would sit on that all night and not do anything is absolutely horrifying," Merv Wiseman told reporters.

The RCMP exacerbated matters by appearing to confirm what may have been a false interpretation of events.

"The first call that was made by the member in Makkovik was to the RCMP's operations support services right here in St. John's. The call to Fire and Emergency Services was not made until the morning of Monday, January 30," outlined RCMP Sergeant Marc Coulombe. "The call to FES could have been done that Sunday night, yes. Would it have amounted to anything? We don't know."

The official timeline compiled by the provincial government and the RCMP does not mention anyone's contacting FES-NL until 7:49 a.m. Monday. On Sunday night, "RCMP Makkovik contacted RCMP Operational Support Services in St. John's requesting air support ... Through discussions with RCMP Makkovik, it was decided to have the search continue in the community."

The matter was raised in the House of Assembly on May 7, 2012, by the acting Leader of the Opposition. "The latest recordings released by DND cast even more doubt on the province's role in the search for young Burton Winters," Dwight

Ball said. "The recordings clearly indicate that a call for assistance left Makkovik on Sunday night, the same evening that Burton went missing. We understand it was the RCMP; there is a coordination and a communication problem here. Given the latest information and the need to revisit existing protocols will you now initiate a public inquiry into the Burton Winters tragedy?"

Dunderdale downplayed the importance of the point and questioned the accuracy of the information contained in the DND tapes: "There is no news that has come to light in the last 24 to 48 hours, Mr. Speaker. As the minister has said, the police in charge of that jurisdiction are the lead in search and rescue activities. They go to the resources available to them—and FES-NL is one of those resources—to request support. FES-NL received the first request for support on the Monday morning, approximately 8:30. That has been confirmed by the minister, by FES-NL's own logs, by the RCMP and by Aliant [the telephone company], Mr. Speaker. There is no news here."

Essentially, the RCMP maintained that they did not ask the province for civilian air support Sunday night, having decided such support would serve little purpose. However, Cpl. Vardy, their officer on the scene, contradicted this and indicated that he *had* spoken with FES-NL that first night. Tapes of a telephone conversation recorded and released by DND tell Vardy's version of events. The conversation between Vardy and SAR coordinator Captain Macdonald at JRCC Halifax took place about 20 minutes before the 444 Griffon was to arrive on the scene on Tuesday, January 31. Macdonald asked Vardy about his plans for the following day: "Tomorrow, are you making any arrangements for helicopter assets through EMO?"

"Everything's going to depend on what happens tonight," Vardy answered.

"Yeah, because they're calling for a great day tomorrow," Macdonald said.

"Yeah, I know," Vardy replied.

"And so non-military resources should have no problem flying," Macdonald continued. "I think EMO should try to make some arrangements tonight so that they're not caught off guard in the morning."

Vardy's voice rose with frustration. "You know what? They won't even do it. I tried to do that the first time and they won't do it. They said: No, no, call us back in the morning. Call us back in the morning."

"That's crazy because you've got 10 or 12 good hours of daylight tomorrow and she should be ready to go at first light."

"Yeah, I know," Vardy said. "You're preaching to the choir ... It's unreal, right?"

"Yeah," Macdonald answered.

"No, we went through that the first night. They never even looked at anything until eight o'clock the next morning and I don't think they were here until after 10 ... no actually that was one o'clock in the afternoon before the helicopter arrived."

Before the call ended, Vardy gave his assessment of Burton's chances: "Right now time is of the essence because if that young fellow, he's on the last, his last legs now ... If he's still alive and we're hoping he is."

Burton may or may not have still been living at that point—but he almost certainly had been for most of Monday, the day after the RCMP either did or did not call FES-NL for help. Whatever was the case, the dispute over this detail suggests that the whole truth is not yet known.

"We all thought the call was going out for air support," the first evening of the search, said Edmunds after the tapes became public. "This raises more questions and furthers calls for an inquiry."

9. The CASARA question

Canada's Civil Air Search and Rescue Association (CASARA) has provided assistance to the Air Force and to other federal, provincial, and territorial agencies during emergency SAR operations throughout the country since 1986.

"When lives are in danger, CASARA volunteers are ready, willing, and able to provide assistance," the association declares on its website. "We are pilots, navigators, spotters, search coordinators, electronic search specialists, radio operators, and administrative staff. We are trained to work as spotters on military aircraft and to carry out searches using light aircraft and ground vehicles ... The mission of the Civil Air Search and Rescue Association is to support Canada's search and rescue program and to promote aviation safety." Funded mainly by DND, CASARA's national office acts as an umbrella group to oversee provincial and territorial member organizations that are, in turn, subdivided into several zones. There are more than 100 zonal organizations across the country. Newfoundland and Labrador had four in 2012—St. John's, Deer Lake, Happy Valley-Goose Bay, and Wabush—but this number has since been reduced to three. Nationally, CASARA has access to at least 375 private aircraft, mostly fixed-wing, and has more than 2,800 volunteers—2,596 of which are certified pilots, navigators, and spotters. The others are involved in support positions on the ground, not only during an emergency but also throughout the year. The qualifications required to be a member of CASARA are quite stringent.

"Pilots are required to have a private licence and at least

200 hours total and 150 hours PIC [Pilot in Command]. Pilots new to CASARA will first be assigned to a trainee navigator's role to learn how to plan search assignments and how to safely and accurately manage the progress of a search flight," states the organization's recruitment literature. "Navigators may be pilots (current or not) or non-pilots with training and/or aptitude for airborne navigation. Non-pilot members with an aptitude and interest in becoming search navigators will be provided the training required to start the search navigation training. All navigators are provided with search navigation training to meet operational requirements."

To be a volunteer navigator a candidate must possess a "private pilot licence (preferred), or considerable air crew experience and/or training; current Licence Validation Certificate or Medical Self-Declaration; 20/20 vision (natural or corrected); good concentration; [and] good VFR chart skills." The spotters, on-site ground team members and other support crew, are all offered a chance to help out in some important way: "If you aren't cut out for flying or hacking your way through the brush, we have other places for you in the CASARA team. Volunteers are always needed to assist with administration and other office duties—a major component of our operations."

The training for spotters, navigators, and pilots is intensive, comprehensive, and frequent—and all carried out by DND, to DND's standards. All volunteers must retrain and upgrade every 18 months to two years to remain in the organization. Brian Bishop, who became president of CASARA Newfoundland and Labrador in 2012, told a SARSCENE conference in 2010 that CASARA does constant "rust proofing" on its volunteers to make sure they are always ready, capable, and up-to-date. Speaking specifically of the St. John's

Zone, he said volunteers were trained in survival, first aid, the use of navigational software, and Military Search Master and CASARA Search Coordinator courses, and were taken on flying exercises to practice what they learn. "The St. John's Zone is committed to providing our volunteers with meaningful and practical training," he said. "We continue to strive to ensure our Zone reflects a diversity of age, experience and expertise in our membership."

CASARA and its volunteers "create an awareness of the Search and Rescue System including defensive flying practices across Canada; provide an array of training equipment such as training Emergency Locator Transmitters (ELTs), ground-to-air radios, computer equipment/programs, telephone pages, Global Positioning Systems (GPS), etc.; provide SAR Coordinators during actual SAR missions and training exercises; provide aircraft and certified crew members such as pilots, navigators and spotters, who can be used effectively on search operations and training missions; provide certified spotters on Canadian Forces aircraft as required; provide rapid response to locate missing aircraft or ELTs in their own areas before Canadian Forces primary SAR aircraft can be airborne ... and provide search headquarters at locations across Canada during actual SAR missions and training exercises."

During a television interview in early 2012, Bishop described how a CASARA team best responds to a call for help: "[I]f we know the night before, if we get a call saying somebody was missing in the night-time, if somebody were to call up and let us know we may have a search on tomorrow, I can have my volunteers here at the airport with all their paperwork done, their maps, their flight planning, the aircraft walked around in preparation to fly and fueled up, we can be airborne at first light. However, if you wait and we get a call halfway through the search, well we've got to get our volunteers rounded up and brought to the airport. It's going to delay things."

By 2010, the St. John's Zone had established liaisons with JRCC, the military, the local Coast Guard Auxiliary, and the St. John's Airport Authority. It had assisted the military on a number of actual searches, providing CASARA aircraft and crews and spotters and other personnel as requested.

CASARA has had as many as 125 volunteer members in Newfoundland and Labrador, but on January 30, 2012, there were 80—six of whom were stationed in Happy Valley-Goose Bay. There were also nine CASARA aircraft available on the island, but none in central Labrador. "The CASARA spotters normally have their own aircraft to go, but Goose Bay is short an aircraft right now," Captain Macdonald of JRCC Halifax was recorded on DND's official tapes telling an unidentified provincial official during the search for Burton Winters. "They have six spotters that are available and they're trained by the military to go out and do these types of searches. They'd be very useful to the guys going out tomorrow."

In May 2012, Bishop confirmed Macdonald's assessment. He told reporters: "The only thing we had available in Goose Bay at that time, we had spotters available who could have assisted with the search. We don't have a CASARA aircraft in Goose Bay at the present time, but we're working on having a charter aircraft and hopefully that will be up and running by late this summer."

Those six highly trained CASARA volunteers, ready and waiting throughout the Makkovik operation, were never asked to help in the search. In fact, the provincial SAR official contacted by JRCC indicated he had never heard of the organization. On January 31, Macdonald had to explain: "Yeah, they're national across Canada. They do a lot of the air searches. They're the equivalent of ground SAR volunteers and they're available to all the EMOs. I'm surprised you don't know about them."

A lack of familiarity with CASARA was not confined to a single FES-NL official. When the public learned that this help had been ignored during the crisis, both the RCMP and the provincial government displayed confusion about how CASARA was supposed to operate.

On May 7 in the House of Assembly, Municipal Affairs Minister Kevin O'Brien pointed out that it had been the RCMP's responsibility to call in CASARA: "Mr. Speaker, as I stated in this House some time ago, the lead agency when it comes to ground search and rescue in this province and elsewhere in Canada is the policing agency that has jurisdiction in that particular region. That policing agency has the sole responsibility of engaging any resources that they might need in regard to supporting them in a ground search and rescue operation. CASARA is one of those and that would be the lead; they would take the lead in regard to engaging those as spotters."

The RCMP had a better handle on the organization—but it was not perfect: "My understanding on CASARA is that it's through DND," said Sgt. Coulombe. "The process to request assistance from CASARA, I couldn't tell you how it works in this province. Whether it's done through Fire and Emergency Services or the RCMP, I'm not aware."

The problem: CASARA didn't work with FES-NL, or with any other Newfoundland and Labrador agency, including the RCMP, which operates under contract with the provincial government. CASARA services are available to DND and to different agencies in the Northwest Territories, Nunavut, and most provinces under the terms of a memorandum of understanding (MOU) that it signed with each political jurisdiction. At the time of the Burton Winters search, CASARA had no such agreement with Newfoundland and Labrador; because of this, only DND could legally call them for help.

"We've had a number of negotiation sessions with the government and their representatives," CASARA's national president John Davidson said in 2012. "We've even sent our legal advisor down to assist. We've been doing this for roughly 20 years trying to get the MOU in place. Nobody seems to have any follow-through on it once we finish the meeting—lots of good promises and lots of good effort put forward, but we never see any results. We're very frustrated."

Bishop, CASARA's provincial president, shared that emotion: "It's frustrating because our volunteers are highly trained and they get frustrated when they know there's somebody out there in trouble and we could help and we don't get a call. We don't have an MOU with the provincial authorities. We've had several meetings going back 20 years with EMO, Fire and Emergency Services, RCMP and the RNC. I've tried on several occasions to get an agreement signed with those groups and if they sign the agreement they could task us directly. DND can only task CASARA for a federal responsibility. This type of search [for Burton] was a provincial responsibility, so that would have to be tasked by RCMP in this case, or the Fire and Emergency Services."

When pressed, Premier Dunderdale implied that CASARA volunteers had not been needed for the Makkovik search; she was vague when asked about the delay in signing an agreement with CASARA: "Mr. Speaker, talks with CASARA have taken place in this province over the last 20 years, both under Conservative administrations and Liberal administrations, and we have not been able to come to a place where we are prepared to sign an MOU. It has just not been possible for a number of reasons."

Outside the House during a media scrum, the premier offered little additional clarity: "There are liability issues and up to this point we haven't been able to settle to the degree that we'd be at a comfort level in terms of the aircraft. Who's flying the plane and so on? These are important issues and we haven't been able to resolve them. However, the talks are continuing."

According to Bishop, all the aircraft are privately owned by CASARA volunteers and were already fully insured. In May 2012 he was not optimistic about seeing any progress toward an MOU: "We're still working on it. They keep saying it's going to happen. Every day, it's getting close. We're getting close, but we're 20 years now waiting."

An agreement was finally announced on June 18, 2013. Government ministers denied that the long-awaited achievement had anything to do with pressure applied since the Burton Winters tragedy.

"This memorandum of understanding is the result of significant work between the Provincial Government and CASARA representatives and formally launches an important agreement that benefits residents of the province," Justice Minister Darin King explained. "CASARA will be another resource available for our provincial police services that coordinate ground search and rescue activities throughout Newfoundland and Labrador."

In the end, all it took was an amendment of the regulations of the Workplace Health, Safety and Compensation Act, which was done in the fall of 2012. The amendment ensured that when CASARA volunteers were called in for a search they would be treated as if they were employees of the Department of Justice, which would give them coverage under the Act.

"The members of NLSARA or CASARA-NL are considered to be in the course of their employment from the time they leave their residence or place of employment to respond to a call as tasked by the Royal Newfoundland Constabulary or the Royal Canadian Mounted Police until they have completed those activities required to respond to that call," the amended regulations state.

CASARA volunteers will never know if their participation could have helped find Burton Winters. "It's a hard question to answer because we didn't get a chance to," said Bishop. "The weather wasn't cooperating and air spotters didn't get a chance to fly in the air, but we do have trained spotters who are trained by the military and we have to keep up to the military standards—we're evaluated every 18 to 24 months. So in an aircraft a trained spotter is an asset, but there's no way I could answer if it could have helped or not."

It matters only that the six Goose Bay spotters should have been given the chance, he continued. They stood ready the whole time, but they were ignored, overlooked by everyone except the JRCC, which, because of a jurisdictional conflict, had no authority to call them in. The six Goose Bay volunteers didn't have their own plane.

According to national president Davidson, they should have been on other aircraft sent to Makkovik that week: "We were there, available. I don't know if that helps, but we've got much in the way of experience and had done this quite considerably … if they had the ability to see the ground, to fly in conditions that were not prohibitive for aircraft flight, then we should have had that good opportunity to assist."

10. Squabbles in the House

The provincial government's actions and reactions in the aftermath of the Makkovik tragedy have not always been consistent. On one hand, it demanded full disclosure by the federal government and has declared the explanations from DND unsatisfactory; on the other hand, it has shown reluctance to disclose the information it does know, unless doing so might aid in its dispute with Ottawa, and it has flatly rejected the idea of launching an independent inquiry itself. In fact, it eventually declared that an inquiry of any kind was unnecessary.

Immediately in the wake of the search, however, a member of the provincial government issued the first official demand for an investigation. Nick McGrath, PC MHA for western Labrador and the Minister of Labrador Affairs, sent a letter to the Minister of National Defence two days after Burton's body was found, exhorting MacKay to look into the matter without delay: "You are no doubt aware of the tragic events which took place in Makkovik this week where a young man lost his life. I understand that the Chief of Defence Staff has initiated an inquiry into the circumstances around the deployment of Search and Rescue aircraft in response to the missing person report. I encourage the federal government to move swiftly in bringing this inquiry to a conclusion, and look forward to learning the outcome. It is vital that this case be reviewed so that lessons can be learned from it and any necessary corrective actions can be implemented. While we do not yet know all the details, the delay in search and rescue response should never have happened and can never happen again."

Four months later, on June 5, 2012, Liberal MHA Yvonne Jones raised the issue of McGrath's letter in the House of Assembly: "Mr. Speaker, the Minister Responsible for Labrador and Aboriginal Affairs stated in a letter to Minister MacKay back in February that it would be vital to do a review of the tragedy in Makkovik so lessons can be learned. Yet he stood in the House of Assembly and stated that a public inquiry would serve no purpose and that it would not change or reveal anything. I ask the minister, why he would do an about-face on this important issue?"

McGrath downplayed his earlier demand, insisting it had been fully satisfied: "There is certainly no about-face. The letter that I wrote Minister MacKay was written two days after the incident happened. The key word in that letter ... was 're-view.' In order to do a review—through our review, we found out that we did not need a public inquiry ... My opinion did not change at all."

McGrath's statement reflected the government's position. Although Dunderdale consistently asked for more answers from the federal government, she stopped short of demanding a full-blown inquiry. Following the two DND press conferences held by Rear Admiral Gardam, Dunderdale had her Minister of Municipal Affairs, Kevin O'Brien, press Ottawa for clarification.

"Upon reviewing this incident, we have questions that we feel would be best answered by the Department of National Defence," O'Brien wrote, in a letter dated February 10, 2012. "I respectfully request your department's assistance in providing us with the clarity we need to determine if any improvements could be made to the process to help ensure the safety and success of future search and rescue operations."

Specifically, O'Brien inquired about the two broken-down Griffon helicopters in Happy Valley-Goose Bay; he wanted assurances that the federal government would ensure the province wasn't short of search and rescue aircraft when the next emergency occurred: "Another issue of concern related to Rear Admiral Gardam's indication that current DND operational protocols mean that DND will not deploy aircraft to assist in ground search and rescue efforts if private aircraft are involved in the search. Given the superior operational abilities of military aircraft and their crews, I believe there needs to be a re-examination of the appropriateness of this protocol. We welcome further discussion on this matter."

More importantly, O'Brien asked why DND had not sent either of its working Cormorant helicopters from Gander to help: "It is our understanding that Cormorant helicopters from Search and Rescue squadron in Gander were available on the morning of Jan. 30, 2012 but that a decision was taken not to deploy these resources in aid of the search effort at Makkovik. I am asking that you provide additional information regarding the availability of those aircraft at that time, the protocols for deployment and the reason(s) why they were not deployed to assist in the search efforts."

To many monitoring this issue, it seemed that the provincial government was asking questions that would justify a full, formal, independent public inquiry by Ottawa or the province, or both. But Dunderdale did not ask for one.

On March 5, Lieutenant-Governor John Crosbie delivered the Speech from the Throne to open the 47th session of the House of Assembly. He interrupted his speech to offer a moment of silence to commemorate Burton and all others lost in similar circumstances. "Sadly, there are times when searches end tragically," Crosbie said. "We as a people have witnessed far too many terrible endings. While each and every tragedy is profoundly felt by the loved ones of those it touches, some grip the hearts of people far and wide. Such has been the impact of the death of 14-year-old Burton Winters in January on the icy coast near Makkovik. Newfoundlanders and Labradorians will never forget his fierce determination to get back home to the ones he loved. We are heartbroken by his loss. We mourn for the many like Burton whose lives have been so tragically cut short."

Liberal leader Dwight Ball stood that day to urge the government to listen to the demands being made by those closest to Burton: "Just last week I had the opportunity to visit Makkovik and meet with the family and friends of this young man. There was one phrase that really struck home to me, and it said: We do not need a southern solution to a northern issue. I am calling on this government, and the Premier, to lobby their federal counterparts to complete a full investigation and commit to establishing a search and rescue centre at 5 Wing Goose Bay, which is a northern solution."

Two days later, Edmunds stood in the provincial House of Assembly to present the request for an inquiry in the form of a private member's motion. He addressed the Speaker of the House on March 7: "The tragedy of young Burton Winters and the legacy he has left us with has created a province-wide mandate to have a review of search and rescue protocols and procedures. It is our responsibility, Mr. Speaker, as representatives in this honourable House to take this mandate forward. Lives will depend on what we have to say and do here today … I cannot say enough about the on-location search and rescue team, about the coordinators, and the ground search and rescue team itself. This part of the search was conducted flawlessly and the participants went well beyond the call of duty in their attempts to rescue Burton and to bring him back to his family … The call, Mr. Speaker, for air support was made late Sunday evening. It was at this point, Mr. Speaker, that the problems in protocol, procedure, and communications began to raise doubt. Whatever transpired, Mr. Speaker—in terms of organization outside of the coordination centre in Makkovik after the call was made—we had no control over … now we question what went wrong and the need to ensure that whatever it was, that it needs to be fixed."

With those words, Edmunds introduced his private member's motion, calling for an inquiry:

> Whereas we have seen many tragedies and many lives lost due to distress at sea and on land, most recently the death of 14-year-old Burton Winters of Makkovik, Labrador; and
>
> Whereas search and rescue were not adequately deployed during the search for Burton Winters; and
>
> Whereas ground search and rescue is a provincial responsibility and was the first point of contact in the Burton Winters tragedy; and
>
> Whereas the Government of Canada has announced the closure of Maritime Search and Rescue Sub-Centre in St. John's, thus impairing efficient and timely coordination of search and rescue services within the province, Therefore be it resolved that this House calls upon the provincial government to consider conducting a full investigation into the Burton Winters tragedy, examining the actions of both federal and provincial agencies and their coordination and communications;
>
> Be it resolved that this House calls upon the provincial government to evaluate search and rescue infrastructure within and available to the province in

case of emergency in order to determine if there is sufficient and stable service;

Be it further resolved that this House calls upon the provincial government and our federal representatives in the House of Commons and the Senate to consider establishing a permanent search and rescue capability at 5 Wing Goose Bay; [and]

Be it [also] further resolved that this House calls on this House of Assembly to establish an all-party committee to make representations to the Government of Canada to rescind the motion and the closure slated for Maritime Search and Rescue Sub-Centre in St. John's, Newfoundland.

The premier responded to Edmunds's motion by stating that an inquiry was unnecessary, but that government members would support a portion of the Liberal motion: "Mr. Speaker, immediately we undertook a review of our own protocols and that process is ongoing. We engaged with the federal government immediately. I asked the federal government to review their protocols. They have undertaken that. Mr. Speaker, we have an important resolution before the House here today and we intend to support that resolution, asking the federal government to strengthen their search and rescue facilities here in this province."

In support of Edmunds's motion, Yvonne Jones presented the first of many petitions to the House. "Your petitioners call upon all members of the House of Assembly to urge the government to do a full investigation into the Burton Winters tragedy and search and rescue in Labrador, and lobby the federal government to establish permanent search and rescue capability at 5 Wing Goose Bay," she stated. Jones also told the House what O'Brien had said at a meeting in Happy Valley-Goose Bay the previous week: "I heard the Minister of Municipal Affairs speak to the investigation or the inquiry that is ongoing through his department. He did commit at that time to table that report when it was finished on search and rescue protocols and what was happening. We look forward to seeing that report, Mr. Speaker."

O'Brien did not address that particular report, noting instead that the province had acted correctly during the search: "We have undertaken and looked at the actions with respect to the incident at Makkovik, and from the protocol of the provincial government's point of view, we responded to the request appropriately." He then proposed an amendment to Edmunds's motion to remove the demand for a "full investigation."

Jones criticized the proposed amendments but agreed, on behalf of her party, to support the amended motion: "[W]e had originally called upon the provincial and federal government to do the investigation and implement the recommendations. What they have said is that we call upon the federal government to work with the provincial government. I do not have a problem with that. I would say to the honourable minister and to his government today that this in no way shape or form allows your government off the hook, either."

Keith Russell, PC MHA for the Lake Melville district, implied that an inquiry was not the best way to get answers, and that Rodney Jacque agreed with him: "When I got to talking with Burton's dad [during a vigil in Happy Valley-Goose Bay, February 23], we came to an agreement on one point at the end of our conversation: It was that this is not a time to make judgments about exactly what happened as Burton got lost, or to look for someone to lay blame upon, but it is time for us to come together as Newfoundlanders and Labradorians and make sure that we work together to do whatever we can to make sure that another family does not have to experience such pain and loss. Though there are details that are still unclear, in conclusion, about exactly what happened, one thing is crystal clear: We all have questions which we would like to have answered. I believe today that with amendments made to the private member's resolution we should get some of those answers."

The vote on Edmunds's motion was eventually called and it, as amended, passed unanimously.

•••

While the debates continued in the House of Assembly, behind the scenes the government secured funding for new search and rescue equipment, the type of equipment that could have helped the ground teams in their search for Burton: $510,000 for 26 thermal imaging cameras and the training to use them. This money went to the Newfoundland and Labrador Search and Rescue Association (NLSARA), which oversees the province's 25 regional GSAR teams. These cameras use infrared light frequency to enable the user to see in the dark or at other times of limited visibility by detecting the heat given off by a person or an object.

"Our government is committed to ensuring first responders have appropriate tools to assist them in the important work they do and to further improve the likelihood of saving lives in difficult situations," O'Brien stated in a March 7 press release. "This new equipment will augment the services provided by ground search and rescue teams throughout Newfoundland and Labrador."

In the same document, NLSARA president Harry Blackmore echoed the minister's sentiments: "These new cameras will be a critical tool for teams to avail of in carrying out their duties in a search effort. We are pleased to partner with the provincial government in acquiring these thermal imaging cameras to build on the assets already available to members." These existing assets included a $75,000 NLSARA annual operating grant from the province and a total of $600,000 over six years to purchase boats and hovercraft for use in search operations.

One thermal imaging camera went to Makkovik, where it was welcomed by SAR coordinator Barry Andersen, who noted, "It has limitations ... It cannot see through objects, or snowbanks, or thick snowstorms. It's not perfect, but it's a great tool."

In the month following Edmunds's private member's motion, Opposition MHAs continued to present petitions to the House of Assembly asking for an inquiry and a SAR base for Labrador; little changed until *The Fifth Estate* broadcast their "Lost on the Ice" episode. The program pushed the issue back into the forefront of public discussion, enflamed the family's passions, and led the Opposition to increase pressure on the provincial government to take more action.

This heightened pressure did not translate into action, but it did spark a squabble in the House over which party was doing the most to lobby federal officials. The dispute ostensibly started with a trip three Liberal MHAs had taken to Ottawa in early April 2012 to meet with federal politicians. Randy Edmunds reported on the Burton Winters Facebook page that he, Yvonne Jones, and Dwight Ball met with officials from the Coast Guard and DND; Jones even met with MacKay. They failed to get a meeting with Labrador MP Peter Penashue, Edmunds wrote, but managed to bring two key issues forward on a federal level:

1) The SAR system failed in the tragedy of young Burton and we have heard every excuse in the book as to why the system failed. Both DND, Fire and Emergency Services, and Coast Guard have no choice but to admit this. Some of these departments have made some indication that they hope to address this in the future.

2) The need for MRSC is vital to our province and the need for it to continue means that lives will be saved. Coast Guard maintains that this centre will close as scheduled with the reason being that advanced communications has made this centre redundant. There are arguments that are contrary to this closure.

The provincial government did not appreciate the Opposition's efforts. Dunderdale accused the three Liberal MHAs of shirking any true efforts to pressure Ottawa and also of "playing politics." She accused Jones specifically of not asking the defence minister for an inquiry when she had the chance—and Jones asked Dunderdale to admit that she had neglected to properly lead the effort to get a better search and rescue system.

"Mr. Speaker, I have to tell you that I find this line of questioning offensive," Dunderdale declared. "We have pressed that case at every opportunity, from the Prime Minister's office, to Minister [Keith] Ashfield's office, to Minister MacKay's office. We have talked to Rear Admiral Gardam, Mr. Speaker, and every other minister ... Every other meeting we go to federally, provincially, it does not matter under what opportunity we have a meeting with federal ministers, Mr. Speaker, we make the case and we are looking for answers because that young man deserves nothing less, nor does his community, nor do people of the province." The spat heated up: Jones asserted that MacKay did not receive the call Dunderdale claimed she had made; Dunderdale then accused Jones of never actually having discussed anything with MacKay, merely posing with him for photographs: "We saw them stood up all over Ottawa, taking their pictures, them and their MPs, Mr. Speaker, outside of Parliament ... I did not see you in any meetings."

Jones could not be deflected: "As the Premier of the province I would encourage her, I would beg her, Mr. Speaker, to go and meet with [MacKay]. We were told that he called you, Premier. You never, ever phoned his office on this particular issue. What we do know now is that the federal government is prepared to be forthcoming with information if the province chooses to do an inquiry into this matter. We know that the province has the jurisdiction to do it. We know that you have the authority to do it ... You are the Premier. Act like one!"

The premier parried with an insult of her own: "If anybody wanted lessons on poor, bad, distasteful politics, you could not have a better lesson than you are getting here today."

Some concrete information did emerge from the exchange. For instance, Jones reminded Dunderdale that her government had exercised the power to unilaterally call for an inquiry in the past, as was her right. Jones quoted from the provincial Public Inquiries Act: "The Lieutenant-Governor in Council may by order establish a commission of inquiry to inquire and report on a matter that the Lieutenant-Governor in Council considers to be of public concern." She cited the 2009 case of the crash of a Cougar helicopter offshore Newfoundland. If the government could inquire into that incident, she said, it could also look into the Burton Winters case.

Dunderdale, however, insisted there was no comparison

between the two events. "I do not have the authority to launch an inquiry into federal government jurisdiction and the deployment of federal government resources. The instances that she refers to all fall under different authorities, Mr. Speaker, like the C-NLOPB [Canada-Newfoundland and Labrador Offshore Petroleum Board] in terms of the Cougar helicopter crash."

The next day the barbs resumed. Dunderdale said she had spoken to MacKay and he had contradicted Jones's account of their meeting: "I had a lengthy conversation with Minister MacKay this morning, and I am interested that ... nowhere in the meeting [between Jones and MacKay]—of all the cry we have had in this House for an inquiry into how the federal government deployed their search and rescue infrastructure, Mr. Speaker, not once did she ask Minister MacKay for a joint inquiry into what happened on January 29 in Labrador."

To which Jones responded tartly: "Mr. Speaker, the Premier probably got her information as confused in this case as she has in every other case that she speaks on. Give her another day and her story will change again. Mr. Speaker, in the conversation that I had with the Minister of National Defence—and unless he denies this or lies about it—this is exactly what the question was and this is exactly what the answer was. I asked him directly: If the Province of Newfoundland and Labrador was to do an inquiry into the Burton Winters tragedy, is there any reason why they would not cooperate and provide the information that was required by the federal government? Mr. Speaker, I challenge the Premier to find where I have said anything different than that."

Dunderdale continued questioning the value of the Liberals' meetings in Ottawa, injecting speculation about whether Ball, the Leader of the Opposition, possessed "extra-sensory perception, or supersonic ears."

Jones's frustration grew: "If I had to start correcting everything the Premier said, I would not get another thing done; I would be all day long correcting the misinformation and all the confusing information that she continues to put out there for the public. I say to the Premier: I had the meeting with Peter MacKay; that is a lot more than you did as Premier of this province, I say to you. In two months you finally made your way to make a phone call!"

Jones next commented on the length of time it had taken provincial authorities to react to the need for an aircraft in the search zone: "According to the log sheets that were obtained from the Department of National Defence, it was 21 hours after the call left Makkovik before the provincial emergency measures organization contacted the joint [rescue] centre for support. It was eight hours later when the joint research centre closed the file and another 23 hours had elapsed before the emergency measures organization of the province requested air support …"

At that point, O'Brien entered the fray: "Mr. Speaker, to hear such false information laid on the floor of this House of Assembly certainly appalls me … the first call that came into Fire and Emergency Services asking for air support was at 8:19 on Monday!"

Edmunds, although he hadn't been recognized by the Speaker, loudly contradicted the minister. "I don't care, I say to the honourable member, if you were there," O'Brien responded. "You ask the RCMP when the first call came into Fire and Emergency Services. Get up on your feet and ask the RCMP when it came. I don't care if you were there … You are playing politics with a tragedy, I say to the honourable member. Appalling!"

11. One grandmother to another

It's safe to say that most of the people, certainly most Labradorians, who followed the debate in the House of Assembly, sided with the Liberals. Many expressed dismay that Randy Edmunds, of all people, would have to face accusations of playing politics. Facebook again provided a forum for many to share their opinions.

"Shame on them [the PCs] and especially Kevin O'Brien for making such a statement when the MHA fights on behalf of the family of young Burton Winters to be accused of playing politics with the tragedy," wrote the MHA's sister, Sharon Edmunds. "I am proud of my brother for doing everything he is doing to not let this issue disappear. He will not be a silent witness like others. He will continue to fight for his people, unlike others. Don't ever give up on this issue, Randy."

Many others published statements of support for Edmunds and his party: "So disgusted with Dunderdale's reproach, who frankly insulted Yvonne Jones and the Liberals during the CBC interview this evening … Jones is asking for a provincial inquiry and so she should." Another: "Randy Edmunds dropped into Tim's earlier. He got quite the cheers/claps from us. We are so proud of all that Randy and Yvonne have done!" And another example: "I was just on VOCM nightline and said that Kevin O'Brien's comments to Yvonne Jones were very disheartening to hear. I said Randy and Yvonne are in no way using this as a political gain … I said Kevin O'Brien needs to back off and think about what he is saying because not only is he throwing those comments at Yvonne and Randy he is also throwing them at the family."

Burton's family thought so too. Spurred on by the venomous exchanges in the House of Assembly, Rodney and Natalie

Jacque wrote an open letter to admonish politicians for their lack of respect: "We are grieving parents who have lost their child in an incredibly dreadful way. Our grieving process has not entirely begun because we are in shock and utter disbelief, not only that Burton is gone, but in the handling of his case during the days he was missing." The letter insisted that the "clear mishandlings" committed by various authorities during the search fully justified an official inquiry: "Personal feelings aside, how are there actual sides to this tragedy? We cannot comprehend how there are opposing views on the matter ... How are political parties arguing over the events when there are obvious errors in the management of Burton's case?"

The Jacques were not surprised by DND's failure to find substantive fault with itself or with its conduct: "It is also not a surprise that a public inquiry is not being pushed by our Premier Kathy Dunderdale to Prime Minister Stephen Harper. There were both federal and provincial misjudgments throughout the handling of Burton's case and this is the reason there has not yet been a public inquiry. Both levels have failed Burton and now there must be a review into the events." Burton's stepmother vowed they would not give up their cause until they found success: "Time was crucial in saving Burton. Yet now that he's passed on, Rod and I will give all our time in pushing for the rights our son never had the chance to receive."

In late April 2012, Dunderdale agreed to meet and speak with a member of Burton's family for the first time. The meeting was arranged for May 11 in St. John's and announced by Charlotte Winters-Fost, Natalie Jacque's mother, on Facebook: "GOOD NEWS TO REPORT: I have a meeting set up with the premier, Ms. Dunderdale, coming up in the near future!"

Word of the meeting, which would also involve Rodney

and Natalie Jacque, brought optimism to those dealing with the province's intransigence. "I'm positive that only good things could come out of this face-to-face meeting," posted Elizabeth Winters-Rice. "That is the hope," Winters-Fost responded. "Burton has gotten lost in the shuffle of 'no ownership' of what went wrong with rescue services in this incident and [I plan] to plead to the Premier to call for an inquiry."

That, however, was not the premier's impression of the meeting's purpose. "Burton Winters' grandmother called and said, 'I want to come talk to you about Burton because he's gotten lost in all this,'" Dunderdale told reporters. "One grandmother to another grandmother, I'm going to sit and talk about Burton."

Two days later, Dunderdale cancelled the meeting. She'd learned that the family wanted to include SAR expert Clarence Peddle, who had loudly criticized the government in February 2012. "The provincial government of Newfoundland and Labrador is now complicit in the failure to get satisfactory answers from the federal government to its own questions regarding the tragedy in Makkovik," Peddle had charged. "It alone is now ultimately responsible for not getting answers to the questions being asked by the family of Burton Winters and the community of Makkovik. It alone is now accountable for not bringing a measure of closure to this tragedy."

Perhaps the premier did not appreciate Peddle's views, or she did not want to face the questions the family, backed by Peddle and his expertise, intended to ask her. She said she cancelled the meeting because she would have been unable to answer technical questions and she didn't want the event to turn into a "mini-inquiry" or media circus. "I'm absolutely happy to meet with Burton Winters' family, but I'm not going to participate in a meeting that is much larger than that, that has a different agenda than what was proposed to me," Dunderdale informed reporters. "In terms of sitting with me and talking with me about Burton and what a fine young man he was and why I don't believe a provincial inquiry won't get us the kind of answers that they believe from the federal government, I'm happy to do that."

Dunderdale suggested that the family meet with someone better able to provide details, such as Municipal Affairs Minister O'Brien. She said they were welcome to bring Peddle along for that. The family declined and expressed their confusion in a letter published in the *Telegram*: "We, the family, felt it was important to us to have Mr. Peddle in attendance and we certainly did not place any conditions on who could be present on the government side … We are unsure of the logic as to why it was acceptable to bring Mr. Peddle to a meeting with Mr. O'Brien, but not to a meeting with Premier Dunderdale."

Or, as Winters-Fost posted on Facebook: "If the family is looking for answers then it is her [Dunderdale's] job to put the family in a place or position where that can happen."

The premier, however, was preoccupied with setting up a meeting with someone else: Prime Minister Harper. "I will discuss what I think is important with the Prime Minister," she told the House of Assembly on May 22. "Unlike the Leader of the Opposition and the Opposition House Leader, when they had an opportunity to speak with Minister MacKay, the Minister of Defence, the minister responsible, they did not take the opportunity to ask for a public inquiry into the deployment of federal resources here on the ground and that is certainly where the question is, Mr. Speaker … I am not satisfied, Mr. Speaker, that the federal government response was appropriate. I understand that they had no dedicated role in it

... that it was up to them whether or not they responded. Past practice says they do it on a humanitarian front, Mr. Speaker. I am not satisfied with the answers I have received."

In fact, Jones *had* asked MacKay about an inquiry, but her question had been about Ottawa's aiding a provincial investigation. That investigation wasn't going to happen—the provincial government consistently stuck to the position that the errors made in the search for Burton Winters were under federal domain and there was no reason the province could, would, or should take the lead and launch an inquiry.

12. Letters between honourable ministers

The first substantive interactions between the federal and provincial governments after the death of Burton Winters occurred in a series of faxes between O'Brien and MacKay. That correspondence remained out of public purview until the province released it to the media to refute the federal government's positions on the matter they'd been discussing.

The first fax, dated February 10, 2012, came from the province. "I am seeking clarity from your department on certain aspects of the operation," insisted O'Brien. He asked questions about the two unserviceable Griffon helicopters in Labrador, the three withheld Cormorants in Gander, and the DND protocol concerning private aircraft. "To be clear, the purpose of this correspondence is not to assign blame. It is about fully understanding the events that transpired January 29-31, 2012, so that we can effect positive change as may be required to ensure the safety of the residents of Newfoundland and Labrador."

MacKay did not seem to construe the letter negatively

and didn't take offence. In fact, when he responded a month later—his return message was faxed to St. John's on March 9—he added a personal touch by crossing out "Minister" in the salutation and writing "Dear Kevin" instead. A "full assessment of all factors" had been completed by February 8, MacKay wrote, explaining at length why the Makkovik operation had been entirely within provincial jurisdiction. The provision of military aircraft to help ground search and rescue should not, for example, be taken for granted: "In general, the use of CF aircraft in GSAR events is reserved for instances when no other option is available or the requirements for the search are beyond the means of provincially secured assets. In this way, the CF can remain postured to respond to our primary SAR mandate, namely aeronautical and maritime SAR assets."

For that reason, DND didn't send the Cormorants to Labrador. Further, the two 444 Griffons were allowed to fall into disrepair because they were not required to be kept in constant readiness for search and rescue missions: "The readiness of primary SAR assets, such as the CH149 Cormorant and CC130 Hercules, is closely monitored ... The CH146 Griffons stationed in Goose Bay are primarily support aircraft. They do not maintain a SAR posture and, as such, their serviceability is not continually reported to the JRCC."

O'Brien was clearly unsatisfied with this response, but he didn't reply until March 27, after "Lost on the Ice" had been aired on CBC: "Recent information by *The Fifth Estate* contradicts information provided by Department of National Defence officials regarding the Department's response to the January 30, 2012 call for aerial search and rescue support," he wrote, without preamble. "Particularly disturbing is that reports suggest DND's operational log contradicts the assertion that the ceiling and visibility on the date in question made it unsafe for helicopters to fly in the area ... We urge you to further investigate and provide us with a formal response. It is imperative that we determine exactly what transpired, resolve any conflicts between statements made by DND officials and the media reports and enable corrective action to be taken. We also request a detailed explanation as to why, if the weather was not the determining factor, the resources in Gander were not tasked."

MacKay's next response to O'Brien arrived only eight days later. He thanked O'Brien for his March 27 letter, but he then contradicted almost every point it made: "Unfortunately, *The Fifth Estate* report contained a number of inaccuracies. In particular, the report incorrectly portrayed the role of the Canadian Forces in the National Search and Rescue Program, the role that weather played in the decision-making process for this incident and the content and conclusions that can be drawn from the SAR incident log." MacKay lectured O'Brien on the various jurisdictions in the country's search and rescue system, essentially repeating the gist of his first letter. He also reiterated that weather had only been one of several factors that had kept DND from sending one of Gander's Cormorants north when it was needed—another, the unserviceable Hercules aircraft in Greenwood—and explained why neither of the Goose Bay helicopters was in working order: "444 Combat Support Squadron stationed in Goose Bay, like all units in the Royal Canadian Air Force, makes every reasonable effort to ensure it has sufficient numbers of serviceable aircraft available to conduct flying operations in the role assigned to it ... On the morning of January 30, one CH146 Griffon was in heavy maintenance while the second aircraft was available for operations. All CF aircraft are complex systems of systems that require many checks and inspections before they fly. These checks and inspections are vital to ensure the safety of our airmen and airwomen. Unfortunately, despite the best efforts of our excellent aircraft maintainers, there are times when the first indication that an aircraft is unserviceable is during these pre-flight inspections. In this instance, during the pre-flight inspection, a mechanical failure was found which did not allow it to launch."

After this exposition, MacKay returned to the issue of the weather, describing how it had hindered the limited Griffon helicopter. The inclement weather hadn't been *in* Makkovik, but on the way there. "In short," he asserted, "if the CH-Griffon had already been in Makkovik on January 30 it would have been able to fly. However, as noted in the CF investigation report, the principal weather-related difficulty was the challenge it would have posed in getting a CH-146 Griffon from Goose Bay to Makkovik in the first place."

Twenty days later O'Brien, still not satisfied, wrote MacKay: "Given the unserviceability of the CH146 Griffons in Goose Bay, we acknowledge that the deployment of other air assets from outside Labrador must take into account many factors including the primary Canadian Forces mandate related to maritime search and rescue. However," he pointed out, "there are two questions that require clarification: According to media reports there were three serviceable Cormorants at 103 Search and Rescue Squadron Gander. Why was the deployment of one of these assets not carried out given that the other Cormorants at Gander would have been available for the primary mandate? It is our understanding that secondary resources were available at Greenwood. Why were those resources not deployed directly to Makkovik?"

According to O'Brien, all credible sources indicated that

weather conditions were adequate to permit military aircraft to fly, and, he added, DND had never backed up its contrary assertions with any proof: "In your letter the emphasis on weather issues is placed on the flight path en route to Makkovik rather than in the Makkovik area itself. By contrast, both the JRCC incident log and the CF report focus on weather conditions in Makkovik rather than en route."

MacKay's next answer, dated May 14, hints at his exasperation: "I am pleased that my earlier correspondence has gone some way toward answering questions you have raised about this tragic incident. I will again address your questions regarding the availability of air assets and the role that weather played in the incident, but there is little new information that I can provide to you that has not already been made available. Nevertheless, I hope this most recent letter will finally put to rest any concerns you may still have."

It didn't. For the third time MacKay lectured O'Brien on SAR's division of jurisdictions, and then he offered another reason why Gander couldn't have spared one of its three helicopters: the way flight crews are organized. "While it is true that there were three serviceable Cormorants stationed in Gander on January 30," MacKay reiterated, "aircraft availability was only one factor considered by the JRCC ... Regardless of the aircraft type in question, the CF SAR standby posture mandates that only a single crew be assigned to each aircraft type at a time. During periods of prolonged SAR activity this posture can be modified to enable scheduled replacement SAR crew to maintain the continuity of SAR service, but this can take up to 12 hours. To be clear, having three serviceable aircraft does not necessarily equate to having more than one SAR asset available with trained crews."

MacKay did not state if this was the case on January 30,

2012, or, specifically, if only one of Gander's helicopters was adequately crewed that day, or if more than one was actually able to fly. Nor did he fully explain why none of Greenwood's secondary SAR assets were sent to Labrador, except that secondary SAR assets are not kept ready, and that Greenwood is far from Makkovik, so private aircraft could get there faster.

On the subject of weather conditions, MacKay elaborated on the risks of flying that day from Goose Bay to the coast in one of the Griffons: "Specifically, the Griffon helicopter is not equipped with an anti-ice capability. The conditions posed a high risk of icing due to the presence of low cloud and snow. This was further aggravated by the presence of mountainous terrain along the 215-kilometer flightpath, which would have forced the Griffon to fly in icing conditions for terrain avoidance."

MacKay concluded with a barb aimed at the provincial government, hinting that it should accept some of the blame: "This continued focus on the role that weather played and the second-guessing of highly trained CF operators is not helpful because it clouds the issue by implying that it is one-dimensional. Unfortunately, in the Burton Winters case, the delay in notification, weather, aircraft availability, and the requirement to maintain the CF primary SAR mandate conspired to deny them [CF SAR crews] that opportunity [to help]."

The words "delay in notification" had been heavily underlined in ink before the latter was faxed to O'Brien. Presumably, MacKay was trying to suggest that this, at least, was caused by provincial authorities, not DND.

O'Brien was not at all pleased. Nor was his boss. They decided it was time to release the correspondence to the media. "Our exchange of letters in the last two months was intended as an effort to answer the questions that have given rise to public concern about the role of Canadian Forces support in the search and rescue activities in Makkovik in January," O'Brien told MacKay in a letter dated May 24. "These letters will be released to the public today."

Although O'Brien acknowledged that MacKay had furnished him with some useful information, he still wasn't satisfied. The government, he said, had determined that one of the Cormorants should have been sent from Gander during the first vital morning of the search—no excuses. He pointed out that, "[w]hile there were reports of poor weather, the weather conditions were changeable and variable. The time it would take to travel from Gander to Makkovik was sufficient time for weather to change. This factor should have been taken into account." In addition, "the one certainty at that critical time was that Burton Winters was missing. This certainty should have received more weight than the weather and the time it would take to mobilize a second crew in Gander to maintain the primary mandate of the Canadian Forces to provide maritime search and rescue."

O'Brien stated that his government believed DND should change its search and rescue protocols around weather and flight crew replacement—that the department should become much more flexible about both factors: "The time to mobilize a second SAR crew should not prevail over the certainty of life at risk in a Ground SAR, especially when the province has exhausted its immediate options for other air support."

Given all that he had written, O'Brien's next statement might have surprised some, and definitely would have disappointed many of Burton's family and supporters: "We [the provincial government] do not believe a public inquiry is necessary. There is ample information in the public domain. We believe a poor judgment call was made, but that does not on its own warrant an inquiry."

Finally, O'Brien refuted any suggestion of provincial culpability, which he called a "deflection of attention" away from what really mattered, saying that it only confused the public: "Firstly, your letter inappropriately emphasizes a 20-hour 'delay' in requesting air support. Secondly, your general observations regarding the provincial Air Emergency Services Program are misplaced ... Finally, your comments about the Civil Aviation Search and Rescue Association's general role in search and rescue are also misplaced. An ample supply of searchers, spotters and VFR aircraft were available for the search for Burton Winters."

If MacKay ever responded to this letter, it was not in the public record as of late 2013.

By the time the province released this correspondence on May 24, 2012, the disagreement between the two levels of government had moved beyond fax machines. The premier had denounced the Minister of Defence for DND's failures on CBC: "I'm certainly at odds with Minister MacKay ... Certainly in terms of search and rescue here in the province, but particularly in terms of the humanitarian response that we looked for [from] them on the search for that young man in Labrador."

13. Ottawa on the sidelines

For months, the provincial PCs stubbornly insisted that any inquiry into the Burton Winters case must be conducted by Ottawa (until O'Brien finally stated they did not believe a public inquiry was necessary)—just as the federal Conservative government maintained that it should be carried out by the province.

Labrador MP Peter Penashue confirmed his government's position in late May 2012 and offered a compromise, of sorts, to the Dunderdale administration, repeating a similar promise made earlier by MacKay. Penashue told reporters that federal officials would not launch a public inquiry, but they would participate in one, as they would be legally required to do.

Dunderdale showed little interest in Penashue's words, even though they seemed to answer her concern that a provincial inquiry would have cooperation from Ottawa. She revealed no intention of changing her government's stance. "Inquiries have served Newfoundlanders and Labradorians well on very specific issues, but not every issue requires an inquiry," she told the House of Assembly. "We are often called, sometimes almost on a daily basis in this House, for an inquiry into something or other."

Penashue was no stranger to the issue of search and rescue. As a Labradorian, he knew as well as anyone the dangers that awaited an unwary traveller in the north, especially in the dead of winter. When he said he understood how dreadful it would be if one of his own children were lost, and when he offered his condolences to Burton's family, his sincerity could not be doubted. As the MP for Labrador, Penashue was expected by his constituents to bring their concerns and demands to the House of Commons. As a member of the Conservative Party, and as a member of Harper's Cabinet, he seemed to lack both the freedom and the will to do it. If Penashue ever asked MacKay to start an inquiry, he failed in his mission. Penashue said nothing further about the proposal, although there were plenty of people who wanted to talk to him about it. Burton's mother, for example, was one of the first to schedule a meeting with him after the tragedy, but

Penashue, she insisted, stood her up. "He could have at least called us to let us know, because right now we're hurting so bad and we just want questions answered," Winters-Rice said two weeks after her son died. "We want someone to tell us something, why they weren't there, why they've got so many excuses. Why can't they just tell us the truth?"

Months later, her husband, Steve Rice, spoke to reporters: "I have been calling Penashue day after day after day, trying to get a meeting set up. So far nothing has happened."

Winters-Rice and Rice weren't the only ones having trouble meeting the Labrador MP.

"I called Peter Penashue, but he never got back to me," said Makkovik Mayor Herb Jacque. "I left messages to Peter Penashue's office, but he never did call back." Jacque, who wanted to speak with the MP about establishing a better and permanent federal search and rescue presence in Labrador, said the secondary assets offered by the 444 Squadron just weren't good enough: "They're not as active as they should be and they're not available when they should be. I think it [a primary SAR unit] should be closer. It should be in Goose Bay."

Few were surprised at how elusive Penashue had become, since he already had a record of dodging the public on the issue of search and rescue. Following Ottawa's 2011 announcement about closing the Marine Rescue Sub-Centre in St. John's, Penashue had slipped out the back entrance of the city's Coast Guard building after helping to celebrate the 50th anniversary of that organization in Newfoundland. Outside the front entrance were about 30 protesters who wanted to tell Penashue that the government should leave the MRSC open. Merv Wiseman was one of the protesters.

"We asked him to please listen to us," Wiseman said. "In

the interest of public safety, please listen to us."

Penashue declined, explaining that he had already made up his mind on the issue and he believed the sub-centre should be shut down to help bring down the federal budgetary deficit. While Penashue also managed to avoid one-on-one meetings with members of Burton Winters's family, he was not so successful at getting away from similar mass demonstrations.

At a protest at Penashue's office on February 10, the protesters did not expect the MP to show up at his office when he did or to stay around when he saw the crowd. Protest organizer Kirk Lethbridge delivered a few words. "As the precious minutes and hours ticked by we hear what had happened, we heard that search and rescue would not come and we heard: If a 14-year-old child is not worth searching for, then what the hell is an emergency?" Lethbridge asked. "It seems that search and rescue is there for a politician who needs to fly around on personal business, but not there for a 14-year-old child in Labrador who is dying of the cold and that is wrong. It's not good enough for us!"

Lethbridge handed the microphone to Penashue. "This is a very difficult situation, I understand that," Penashue began. "I am a father and I am a grandfather so I completely understand the situation in terms of what it would feel like for me. My sympathies and my condolences go out to the family."

For the next 40 minutes or so, the MP listened as his constituents spoke about improving the search and rescue system in Labrador.

"This is about authority, and protocol has to change immediately," said Gary Mitchell, who had grown up in Makkovik and was then serving as Ordinary Member for the Lake Melville area in the Nunatsiavut Assembly. "I am not giving up. We are going to fight for what we believe in. We are frustrated. It's time for action for SAR to change their protocol and give Labrador some recognition and a permanent SAR base here in Labrador."

Before the meeting ended and Penashue left, he promised to bring his constituents' concerns to Ottawa and that he'd participate in another meeting before too long. Lethbridge seemed satisfied. "I think that 'people power' today put some pressure where it needs to be," he explained. "I think that Minister Penashue and Keith Russell are both aware of what needs to be done and the ball is in their court now to push this issue to what we are looking for, which is to have a search and rescue helicopter stationed permanently in Labrador. They know what we want and we've given them the leeway to do whatever it takes in a reasonable amount of time to come back to us and we will have another meeting at that time."

That next meeting took place exactly four weeks later, on March 9, following MacKay's announcement that DND had changed an unwritten protocol and would now take the initiative to call back anyone looking for help, rather than waiting for a second plea. About 80 people gathered in the Labrador Friendship Centre in Happy Valley-Goose Bay to talk with Penashue. However, Penashue had little interest in hearing complaints. The country's search and rescue system was fine as it was, he insisted, and nothing needed rearranging. "All of the assets that are available have all been assigned in the appropriate places to provide the best service we can," he told his listeners. "So if you're looking for an answer from me that there will be a primary SAR similar to what they have in Gander, the answer is 'no.'"

That didn't make sense to most of the people in the room. "What's happening now, whoever is responsible for it, it's not working," responded local activist Petrina Beals. "We have a 5

Wing base, a full military base there, people who are equipped to do this ... We've got a base, why not that base be responsible for SAR in Labrador?"

Gary Mitchell also spoke: "After all Burton Winters went through out on that ice, after all the vigils we had to honour and respect his courage and inspiration in his ordeal to survive, all the rallies we had for SAR in Labrador and all the people coming together, I think this is an insult to the people of Labrador: that Peter MacKay is going to make one small rule change in regards to, 'Don't call us, we'll call you.' I think there should be a lot more than that. I think there should be equipment based on Labrador soil."

One of the last people Penashue heard from in the public meeting was Burton's maternal grandmother, Edna Winters. Her idea of how to improve the system was a novel one: "I think with SAR services it needs to be looked at as an essential service just the same as clinical health and any other public service that's out there. Unfortunately it had to be a family member we lost to bring this out to [a] head. I think the way that they are responding, it's a lot of 'their responsibility, our responsibility.' It's all of the governments' responsibility. I think that they need to come together and work together better."

Whether Penashue or any of his Cabinet colleagues ever took that advice to heart is uncertain. At any rate, there has been no sign it was put into practice. The two levels of government did not get any closer on the issue. The province made whatever efforts and charges it deemed necessary, and the federal government did the same. The only point of agreement was that there be no public inquiry into the tragedy.

More demonstrations took place in front of Penashue's constituency office, but he did not meet with the protesters again. Any statements he gave generally repeated his government's version of the events and sometimes pointed out the improvements that Ottawa was making, including the "new" call-back protocol. In April, however, came the announcement of a new helicopter and what appeared to be a substantial new investment in SAR.

The helicopter was a third Griffon for the 444 Squadron at the Goose Bay air base. Instead of providing Labrador with a primary SAR unit, as Burton's family and others had been demanding, the Conservative government added one additional secondary resource. "A third Griffon aircraft at 5 Wing Goose Bay will have an immediate positive impact on the operational readiness of the base and provide flexibility to decision-makers on the use of Canadian Forces assets in the region," MacKay announced in a news release on April 11. "This helicopter represents another resource that can contribute to Canada's search and rescue system in support of primary responders in this region."

Penashue described the third aircraft as the answer to everyone's demands: "That will bring the number of helicopters from two to three and that will give full services to 5 Wing Goose Bay for their needs, as well [as] for any request for ground search and rescue. Our glass is half full, not half empty. We have to remain positive about the story because I think this is excellent news for Labrador."

Base Commander Lieutenant-Colonel Andrew Fleming concurred: "Adding to the complement of Griffons at 5 Wing Goose Bay will greatly improve the squadron's ability to conduct and maintain aircrew training, proficiency and experience."

The next DND announcement came on April 15: Ottawa was "investing" $8.1 million into "search and rescue prevention and response." This money was to go to the National

Search and Rescue Secretariat to be spent on tools and equipment, on exercises involving the Forces, the Coast Guard, and Parks Canada, and on developing training standards, computerized training systems, and outreach programs. "This money will support projects that build search and rescue capacity and strengthens the response of search and rescue," MacKay outlined.

"Projects funded under the Search and Rescue New Initiatives Fund support the priorities of the National Search and Rescue Program," explained Geraldine Underdown, the Secretariat's executive director. "They contribute to building capacity for the National Search and Rescue Program in such areas as search and rescue prevention, response, technology, sustainability, interoperability and the North."

The announcements of the new helicopter and $8.1 million were immediately welcomed by the general public, but criticisms soon emerged—the primary being that neither the Griffon nor the $8.1 million was actually new. The helicopter, it turned out, had already been stationed at 5 Wing Goose Bay but reassigned when needed elsewhere while the Canadian Forces were engaged in Afghanistan. Nevertheless, Burton's family welcomed its return to Labrador. "They were quite happy with movement being made, in terms of resources being made available at 5 Wing Goose Bay," Edmunds said. It wasn't the full SAR centre Labradorians were demanding, "but this certainly is a step in the right direction."

Similarly, the multi-million dollar grant to the Secretariat was exactly what DND provided to its semi-autonomous offshoot every year. "The New Initiatives Fund is an annual $8.1-million allocation for research, studies, etc., administered by the National Search and Rescue Secretariat," Newfoundland MP Jack Harris wrote in an email to the media.

"The annual budget has been $8.1-million for years ... Not new money." MacKay's office confirmed this description. "A new merit list is submitted each year to support search and rescue initiatives," according to spokesperson Jay Paxton.

Newfoundland MP Liberal Judy Foote called the announcements "misleading and disingenuous," considering all the cuts that search and rescue were suffering. "On the one hand, government is closing down vital life-saving centres, while on the other they are investing into other components of the search and rescue operation. This most recent funding announcement, along with the third helicopter for Goose Bay, which is not a new helicopter for Labrador, but the return of the one that was taken and placed in another province ... is merely a political tactic to divert the public's attention from the tragic loss of Burton Winters, the unjustified closure of the MRSC, and the whole F-35 [an airplane DND wanted to buy] fiasco."

Little has happened since the $8.1 million announcement on the federal level. Even in the House of Commons, where Opposition MPs frequently raised the issue of the country's search and rescue system, government members have generally confined themselves to repeating how they changed one protocol and provided an extra aircraft to Labrador. Beyond that, they have simply rejected Opposition proposals, like the private member's motion Harris submitted on April 30, 2012, to institute a maximum 30-minute response time for federal SAR units.

Conservative MP Chris Alexander, Parliamentary Secretary to the Minister of National Defence, spoke against that motion: "On the premises, we do not agree that Canada lags behind international search and rescue norms ... We also do not think it is the place of this House, this member, or other members to determine what the actual response times of the Canadian Forces, or any other body, ought to be on these matters. The House has never set those standards in the past."

In April 2013, the Auditor General released a scathing report about the deficiencies of the national search and rescue system. Its findings infuriated the public but did little to shake the government out of its complacent confidence that little needed changing.

14. The Auditor General reports

According to the Auditor General of Canada, the country's government-vaunted "second-to-none" search and rescue system is, at best, merely adequate and, at worst, is "near the breaking point." Its office released the periodic performance audit report on Federal Search and Rescue Activities after an investigation covering the period between April 2007 and December 2012. "We examined whether federal organizations adequately oversee search and rescue activities, are ready to respond to SAR incidents, and have implemented prevention activities to reduce the number and severity of SAR incidents," the report states.

The audit team had looked into DND, the Canadian Coast Guard, Fisheries and Oceans, Transport Canada, and the National Search and Rescue Secretariat but not into any provincial, territorial, or municipal organizations. It found numerous deficiencies in response times, standards of readiness, training, staffing, aircraft, equipment, information management, communications, and cooperation with other levels of government.

It discovered, for example, that while the Coast Guard met its 30-minute response-time standard 96 per cent of the time, the Canadian Forces only managed their more lax standard—a 30-minute response time from 8 a.m. to 4 p.m. on weekdays and two hours during evenings and weekends—85 per cent of the time. According to the report, "We also found that the Canadian Forces does not regularly review whether its states of readiness are still appropriate to meet expected needs. Furthermore, current readiness standards were set using the resources available rather than a needs analysis. If standards continue to be based on available resources and this capacity declines, this will result in a reduction of readiness standards and service levels." In other words, more budget cuts would mean poorer service.

Both the Air Force and the Coast Guard were experiencing personnel shortages that hampered their ability to meet their responsibilities, the audit found, and both were finding it difficult to keep their staff properly trained: "The number of crew members per aircraft is at the minimum requirement that the air force has set for itself and it is losing experienced people. While SAR crews conduct their missions with available resources, the loss of experienced personnel results in more pressure placed upon less experienced air crew to perform supervision operations, and training. It also makes it more difficult for air crews to be scheduled for operations and still obtain necessary training, as well as professional development."

The Auditor General audit found the same situation in the JRCCs in Victoria, Trenton, and Halifax, where the air controllers and their assistants were regularly fewer in number: "The required number of controllers was established more than 30 years ago. However, the total number of controllers has decreased, information technology has become more complex and SAR responsibilities have increased." As a result, everyone in JRCC has less support, less time for training, fewer leaves, and more work—compounded by the closure of the Marine Rescue Sub-Centres. According to the audit, "[t]he MRSCs reduced the JRCC's workload in areas of high marine activity. They have been responsible for coordinating marine SAR activities within their respective areas and working with the aeronautical coordinator at their JRCC to provide assistance to air search and rescue."

All the aircraft available for search and rescue were deficient: the Cormorants had maintenance issues, the Griffon was unsuited for long-distance northern missions, and the Hercules was too old and in frequent need of hard-to-find spare parts. The audit revealed that "[t]wo aircraft are in extensive maintenance at any time and it takes all 11 remaining SAR Hercules airplanes to maintain SAR operations. Consequently, the Air Force has little flexibility to meet operational demand and must from time to time call upon other aircraft."

The system's computerized Search and Rescue Mission Management System, which is used to coordinate all efforts and provide vital data during an emergency, was antiquated and near collapse, the audit found. It is not scheduled to be replaced until 2016.

While the audit established that the $8.1 million provided for the SAR New Initiatives Fund was spent properly every year, it ultimately seemed to serve no purpose: "program recipients produced reports, but the NSS did not use these reports to determine whether projects have the potential to reduce the number and severity of SAR incidents, nor were they used to improve future prevention programs."

Finally, the audit found that Canada lacks a national SAR

policy framework that had been identified as being necessary in 1976, and more recently by the 1985 Royal Commission on the Ocean Ranger Marine Disaster: "In spite of the many reports and recommendations for a national SAR policy, we found that there is still no such policy nor an overall federal policy, planning framework, clear statement of expectations for federal SAR services, or ability to measure overall federal SAR effectiveness."

The report concluded that the Air Force and Coast Guard were performing their SAR tasks adequately, but that, without action, this would likely change for the worse: "significant improvements are needed if they are to continue to adequately respond and provide the necessary personnel, equipment and information systems to deliver SAR activities effectively."

In the House of Commons the day after this report was released, the Defence Minister was contrite. "That is not good enough," MacKay conceded. "We recognize that. We accept his recommendations. In fact, with these recommendations, we have already begun work on the issues. We have already begun working with other stakeholders, including other departments. We will continue to do so. We will be assessing our search and rescue governance structure at all federal levels, as well as working with the other jurisdictions to ensure that search and rescue continues to improve for this country."

The Opposition, however, hailed the Auditor General's report as a timely expose and an explanation for the government's reluctance to explore the failings in the search for Burton Winters. "Mr. Speaker, the Auditor General's report spells out significant challenges facing Canada's search and rescue," Judy Foote said on April 30, 2013. "It is painfully clear why the Conservative government refused to call a public inquiry into the tragic death of a 14-year-old Burton Winters, despite repeated requests. Given the Auditor General's report, it is also painfully clear why the Minister of National Defence should not use search and rescue as a limousine service from a fishing camp."

MacKay again promised improvements: "The tragedy of Burton Winters is something we all remember. Our thoughts and prayers are with his family, but let us not lose sight of the efforts that are being made, that are being undertaken."

Two days later, MacKay announced that his department would allow the JRCCs to alter their hours of operation slightly—the 30-minute response time window could start an hour earlier every weekday, or end an hour later (but not both) rather than being confined to the hours between 8 a.m. and 4 p.m.

epilogue
A good man gone and other tragedies

In early May 2013, four men travelled from Sheshatshiu in central Labrador to go bird hunting on Park Lake, a large, shallow body of water on the interior plateau south of the snow-capped Mealy Mountains. In May, Labrador is warming up but can still be as cold as winter. Most of the ice is gone from the rivers and ponds, but the water remains frigid. One of the four men was Joseph Riche. Riche had served as the Grand Chief of the Innu Nation, an organization that has been negotiating land rights for Labrador Innu since the 1970s, from 2010 until 2012. Riche was well liked and many felt he had a long career in public life ahead of him.

Riche's visit to Park Lake with the three other hunters should have been a routine trip—his family had hunted in the area for more generations than anyone need count. But the expedition did not go as expected. On the afternoon of May 8, a member of Riche's party became stranded on an island, suffering a diabetes-related crisis. Riche was paddling toward the island to help when he ran into trouble of his own. His canoe capsized, pitching him into the lake. The two remaining capable hunters looked for Riche in the water as long as daylight let them, then they made a distress call to the RCMP in Happy Valley-Goose Bay, about 80 kilometres to the northwest.

The RCMP received that call at 8:15 p.m. Thirty-seven minutes later, the police called FES-NL in St. John's; four minutes later that organization called the JRCC in Halifax. FES-NL could not immediately send a civilian helicopter to Park Lake because none of those that could be used by the province was equipped to search after nightfall. Provincial officials asked DND to send its nearest aircraft. DND did so—in a manner of speaking.

The Canadian Forces had three Griffon helicopters stationed at 5 Wing Goose Bay (each capable of reaching Park Lake in about 40 minutes), but none was in working order, so the JRCC tasked a Cormorant out of Gander. Since it was a weekday evening, the Cormorant was required to be in the air within two hours—instead of the 30-minute response time required for calls received before 4 p.m.—but the crew was airborne 64 minutes after JRCC was alerted. Gander is a significant distance from Park Lake, and the Cormorant took two hours and 15 minutes to get to the search zone.

No matter what time the helicopter arrived at the lake, it is likely that nothing could have been done to save the Innu leader's life, and searchers were able to recover his body just after noon the following day. Nevertheless, the question arose: why did DND take so long to respond, when a military helicopter could have been there much sooner? Asked why none of the 444 Squadron's Griffons (one of which was the "new" helicopter that had been returned to Labrador in 2012 with great political fanfare) had been able to respond to the emergency, 5 Wing's public affairs officer gave two reasons: first, that search and rescue is not the 444's primary function, and second, that all three helicopters had been in the middle of either extensive maintenance or vital repairs. Captain Dave Bowen said one of the helicopters would usually have been able to fly, but the squadron's mechanics were waiting for a new windshield to be shipped from Calgary.

"A cracked windshield, it doesn't sound that big of a deal, but … it's actually part of the structural integrity of the aircraft," Bowen said. "It's not a quick fix, either. Once they get the replacement part it takes three days for the glue to cure before they can actually fly the aircraft."

While the circumstances of Joseph Riche's drowning and the loss of Burton Winters differ considerably, there are enough similarities—including the lack of readiness of vital equipment and the resulting widespread public mourning—that the military's official spokesman drew parallels.

"I was here for Burton Winters and it's the same level of tragedy. Everybody here at the squadron, they train night and day to do search and rescue … and they feel for every mission," Bowen empathized. "The positive side to this is that the primary SAR unit for this area did respond. They got the Cormorant up here and they ended up doing a medical evacuation as well [the stranded hunter was flown to hospital and made a full recovery]. That in itself is a small ray of sunshine in a really dark day."

That ray was not bright enough for many concerned people. Park Lake is outside of MHA Randy Edmunds's district, but he was forthcoming with his thoughts: "To have to wait for an aircraft to come from Gander, it just blows my mind. I'm just so disappointed … to learn that all three [Griffons] were down was certainly overwhelming."

The Canadian Forces treated the 444's state of mechanical unreadiness as unavoidable and isolated happenstance. Major James Simiana, an Ottawa-based spokesman for the Canadian Forces, responded to criticism by pointing out that everything had gone according to protocol. The federal search and rescue system sent the closest primary resource to help—in this instance, the Cormorant from Gander. In any case, he added, the Goose Bay Griffons only fill a support role: "When required, we do engage as many assets as we can possibly bring to bear to hopefully provide a satisfactory outcome … the unfortunate reality is that reality does intervene and those aircraft were not available to provide support to the dedicated primary assets that did in fact respond."

A similar response had been given by DND almost a year earlier, five months after Burton Winters died—but on that occasion, all of Gander's Cormorants happened to be out of service. On July 5, 2012, help was urgently needed to rescue a man who had fallen off a cliff beside the Exploits River near Grand Falls-Windsor. The man had broken his leg and was lying in a ravine. Provincial authorities called to arrange an emergency airlift, but none of the three Gander-based Cormorants was operational. A private helicopter company had to be called in to help, sparking fresh criticism about DND's usefulness.

"It's a dangerous precedent that they're setting by having a private company do SAR for us when we have our own Canadian Armed Forces out here not properly armed to do the job," NDP MHA George Murphy said on July 6, 2012. "If the Cougar Helicopter was tasked yesterday to do its own job, we would not have had anything in this province, in Newfoundland and Labrador, beside a couple of Griffons in Labrador, to respond to anything that would have been of a crisis nature anywhere else in the province."

Faulty equipment may not be the worst of DND's issues when it comes to promptly joining a search and rescue operation. The confused state of communications between the separate agencies and different levels of government has proven an equal or greater impediment. The military has blamed blind chance for some of the less-than-stellar missions they've been involved in, but they are not able to so easily dismiss what happened during several other searches, successful and unsuccessful, that took place in and around Newfoundland and Labrador both before and after Burton Winters went missing. Retired SAR specialist Clarence Peddle doesn't even count the Joseph Riche tragedy among the problematic missions—but he says he knows of at least six other incidents in Labrador alone when the SAR system malfunctioned because DND didn't respond as it should have, although the fault was not always DND's.

"There have been at least half a dozen cases ... where protocol was not followed," Peddle related. "These were people who would quite often have been subject to a full [Canadian Forces] search and rescue from the beginning, but they were not brought in until much later."

Peddle's primary example is the case of a 13-year-old Natuashish girl who went missing for several days in the autumn of 2011: "She stole a canoe and tried to make her way to see her boyfriend. There was little attention brought to the case because she survived."

The girl was trying to get to Nain, an impossibly long trip for her to attempt, especially as she had no paddle and used a stick instead. She got as far as an island several kilometres from Natuashish, where she became stranded. She survived by eating berries and huddling under trees for shelter until she was eventually rescued.

"Her feet were frostbitten," the girl's aunt told reporters. "She was really thirsty and really hungry and the tips of her fingers were frostbitten, too. She was able to talk and tell me some of the things she did on that island ... She [had] packed a bag and took that in the canoe with her. So she was able to change clothes when she was wet, but she was on her last change of clothes and was wearing a bright orange top as a skirt, and that's the thing that the search and rescue crew saw when they found her."

Authorities had not treated her disappearance seriously enough, Peddle said, and their attitude delayed her rescue by several days. Their main mistake was that, even after the canoe's owner reported on October 1 that his boat had been stolen, no one connected its disappearance with the missing girl. Because she had had a fight with her mother, the assumption was that the girl had not left Natuashish at all but was hiding in another residence in the community. The girl did not become a federal responsibility until it became clear that she'd taken to sea in the stolen canoe, and it was almost too late. Because of that delay, she wasn't found until four days after the canoe was reported as missing; she may have been on the island longer than that. RCMP told reporters that they didn't know when she had actually taken the canoe.

"One more night and she would have been gone," Peddle said. "Very little has been said about the effort put into finding the young girl ... She was on the open ocean, so it was a federal search and rescue, but they were never made aware of it until later, but even when they were made aware they were somewhat dismissive of it."

Peddle and Edmunds cite two other emergencies which have bolstered demands for improvements to search and rescue. Peddle says he saw the same half-hearted response that endangered the Natuashish girl in the 2005 sinking of the *Melina & Keith II*, a 19-metre fishing boat. According to the investigation report by the Transportation Safety Board of Canada, the boat left Catalina to fish turbot and shrimp with eight men on board: seven crewmembers and one fisheries observer. Three days into the trip, the boat had more than 30,000 kilograms of catch. While the crew was pulling in the nets to load more, the vessel listed to port, recovered, but then slowly turned over. The eight men had time to save themselves by walking and climbing up over the rolling hull, finding temporary safety perched on the keel as the *Melina & Keith II* stabilized upside down. Their relative safety, however, did not last long. The vessel sank two hours later with no help in sight. By the time a helicopter arrived to pluck them from the sea, four of the men had drowned.

"It wasn't necessary for [them] to die," lawyer David Bussey said in 2007. Bussey represented the father of victim Justin Ralph, a 21-year-old crewmember, in a suit against various federal government departments and agencies. "[Justin] actually was hanging on to the bottom of the vessel for a couple of hours and should have been rescued, but he wasn't."

The four men died, according to the Transportation Safety Board, because the search and rescue authorities showed no urgency in responding to the Emergency Position-Indicating Radio Beacon (EPIRB) that started broadcasting a distress signal shortly after the *Melina & Keith II* rolled over. A satellite picked up the beacon at 3:32 p.m. and alerted the Canadian Mission Control Centre in Trenton, Ontario, which passed the information on to the JRCC in Halifax, which in turn informed the MRSC in St. John's. That all happened in three minutes, but it was still more than three and a half hours before help reached the crew. For the first 51 minutes, all the SAR coordinator on duty did was attempt to call the two contact numbers on file for the boat. He never got a response.

"MRSC St. John's made several attempts to contact the owner by calling the two telephone numbers listed on the

EPIRB data sheet, both of which produced a 'no answer' signal," Clarence Peddle wrote in an official report a month after the incident. "At this time there were several actions that could and would normally be made to respond to a distress incident and to obtain more information. There was however no distress atmosphere, but instead an atmosphere associated with an EPIRB false alarm, no doubt triggered by the high incidence of electronic false alarms."

Edmunds evokes the case of the *Ryan's Commander*, a 20-metre fishing boat that capsized about a year before the *Melina & Keith II*. The *Ryan's Commander* had six men on board and nothing in its hold—the crew had just unloaded a load of shrimp in Bay de Verde—when it turned over near Cape Bonavista. Everyone on board managed to get in a life raft before the boat fully capsized.

They may have remained safe, except search and rescue did not arrive until the life raft had drifted dangerously close to shore. Pressed for time and space and lacking the use of two fully functioning hoists—one could not be used in the heavy winds that were being experienced—the crew of the Cormorant helicopter sent from Gander were not able to prevent two of the fishermen from perishing in waves at the base of a cliff.

According to the report of the Transportation Safety Board, all the proper search and rescue protocols had been followed. Critics say that if that's true, the protocols are flawed.

"The *Ryan's Commander* capsized outside of regular Department of National Defence ... working hours and the primary SAR air resources at Canadian Forces Base Gander were operating on a maximum of two-hour response time," the report reads. "The response helicopter departed for the scene within the prescribed response time. However, by then, the life raft had drifted closer to shore. The minimum difference in response time between the maximum 30-minute response in use during working hours and the actual deployment during the occurrence amounts to 30 minutes. While it is difficult to determine whether the outcome of this occurrence would have been different had the SAR helicopter arrived 30 minutes earlier, it is possible that the effect of the winch and hook problems on the rescue would not have been as critical and that, with the life raft further offshore, the SAR crew would not have been hampered by the close proximity of the shore and cliffs."

These tragic examples show that the failures in the Burton Winters search were not isolated. The Burton Winters case can be set apart from the others because the delayed and inadequate response wasn't just caused by equipment failure or jurisdictional confusion or a lax appreciation of the urgency of the situation—but by all three factors, and possibly more.

Furthermore, these varied examples show that even if Burton hadn't gone missing, there is cause to call an inquiry into the federal and provincial search and rescue system. With the Burton Winters case in the mix, the call becomes overwhelmingly loud.

"I don't know what it will take [to get an inquiry]," said Edmunds, adding that the provincial Liberal Party will continue to push to see it happen. "I think our position on SAR is that it will encompass a number of incidents that have raised questions and not just one of the tragedies that have occurred that raise questions about SAR."

As for whether an inquiry would be of value, one need only reflect on what Newfoundland's Municipal Affairs Minister wrote to the federal Minister of National Defence on May 24, 2012, about not needing one. "We do not believe a public inquiry is necessary," O'Brien informed MacKay. "There is ample information in the public domain. We believe a poor judgment call was made, but that does not on its own warrant an inquiry."

The "poor judgment call" to which O'Brien referred was DND's decision not to send a Cormorant helicopter from Gander to Makkovik on Monday morning. If that had indeed been the only factor that had led to Burton's not being rescued in time—or if it had been the only time a poor SAR-related judgment call had made a bad situation worse—then O'Brien might have had a point. It is glaringly obvious that that was not the case.

Not only was the "poor judgment call" *not* isolated to the Burton Winters tragedy, but it also appears that more than one error was made during that emergency. Peddle pointed out at least two more: why didn't the RCMP (or FES-NL— that also remains unclear) arrange for any type of aircraft until a full night after Corporal Kimball Vardy asked for one? And who let the contracted Universal helicopter go back to Happy Valley-Goose Bay early Monday afternoon when there was still a chance Burton was alive? That happened after many feared Burton had driven his snowmobile into the icy waters.

"Why was that helicopter sent home?" Peddle continued. "That question has never been answered … I don't know who stood that helicopter down at 2 p.m. when they had a beautiful day and they could have found the snowmobile."

Federal and provincial authorities failed to have aircraft on the scene, which would have almost certainly, according to Peddle, have brought the search to a successful conclusion: "I'm convinced they should have found Burton Winters Monday afternoon and they would have found him alive."

Instead, both levels of government left it up to the ground

crews to look for the lost teen. Able to cover just a fraction of the territory a helicopter could scout, they never even came close to discovering Burton's Tundra. By nightfall, the youth's last chance was gone.

Peddle also said that the handling of communications during the emergency has to be examined in more detail. The information that is known, incomplete as it is, shows that the system slowed down the response. Peddle insisted that Vardy, the man coordinating the on-site search, should have been able to contact the JRCC in Halifax directly—instead, he had to go through the St. John's RCMP and then FES-NL. The FES-NL coordinator on duty, Paul Peddle (no relation), became the established middleman between Makkovik and Halifax.

"That was the most stupid part of this operation," Clarence Peddle said. "Why did Corporal Vardy have to go to Peddle? He should not have been anything more than a stamp of approval. Peddle should not have been in the equation." Clarence Peddle lists the inaccurate information that was relayed to JRCC in this roundabout fashion: false statements about what the search teams were finding and incorrect locations for both the snowmobile and the body. "He [Paul Peddle] didn't know any of the coordinates. When they found the Ski-Doo, he called up and said they found the Ski-Doo under water."

Given all of these details, Peddle was quite clear about what he thought should happen and why. Regardless of when an inquiry is launched, he said, it will be necessary and valuable: "It's important to realize the system failed because of individual mistakes and a flawed system that allowed them to happen—a system that is still in place. The passage of time should never exhaust the requirement for an inquiry. People should not be let off the hook because five years have gone by. People should be asked these questions in a courtroom setting. There are people who have never been identified who played a part in this."

Peddle, however, sees no chance that either of the Conservative governments in St. John's or Ottawa will order an inquiry unless forced to do so. The politicians are against an inquiry because it would probably prove that the governments themselves bear responsibility for Burton's death. "Nobody," he pointed out, "wants to admit they made mistakes. That's why they come up with the convoluted explanations."

Peddle is putting his hopes on the provincial Liberal Party, since its members have promised to hold an inquiry the next time they are in power, which could happen in the 2015 general election. He does not think the federal Liberal Party will do the same.

However, Labrador MP Liberal Yvonne Jones, who defeated Peter Penashue in a May 2013 by-election and was named the party's Search and Rescue Critic, suggested otherwise. Jones said an inquiry must happen one way or another—launched by either level of government, or conducted jointly.

"It's never too late to do an inquiry," Jones said. "I will always push for one. If ever I'm in a position where I can do this as part of the government, I will do it."

For now, the final words on the subject should be those spoken by Burton's mother days after she lost her son: "We want someone to tell us something, why they weren't there, why they've got so many excuses. Why can't they just tell us the truth?"

acknowledgements

This book could not have been researched and written without help from numerous people and without drawing from the work of dozens of reporters and writers. I would like to thank everyone who consented to an interview for this book: Barry Andersen, Randy Edmunds, Willie Flowers, Anna Jacque, Herb Jacque, Yvonne Jones, Enid MacNeil, Dalton Manak, Elizabeth Mitchell, Clarence Peddle, Terry Rice, and Jacqueline Winters. I would like to credit many journalists whose work I referred to, including Peter Cowan, Gillian Findley, Colleen Connors, Tony Dawson, Rob Antle, and Lee Pitts with CBC; Derek Montague, Alicia Elson, Terri Saunders, Jamie Lewis, Steve Bartlett, Rosie Gillingham, Colin MacLean, Dave Bartlett, Andrew Robinson, Justin Brake, James McLeod, Ashley Fitzpatrick, and Daniel MacEachern with Transcontinental Media; Stephanie Levitz, Sue Bailey, and Aly Thomson with the Canadian Press; Greg Burchell and Lee Berthiaume with Postmedia; Samantha Bayard with Straight Good News; Bill Curry and Steven Chase with the *Globe and Mail*; Alex Ballingall with *Maclean's*; David Pugliese with the *Ottawa Citizen*; and Ossie Michelin and Taryn Della of the Aboriginal People's Television Network.

As well, I would like to thank David Lethbridge, Herb Brown, Dawna Lee, Greg Pastichi, Marharla White, James Johnson, and the library of the Labrador Community College in Happy Valley-Goose Bay for providing me the time and space to work on the manuscript. Also, thanks to Gavin Will of Boulder Publications for giving me the chance to write the book, Stephanie Porter for her excellent editing, and Iona Bulgin for her diligent copy-editing.

about the author

Michael Friis Johansen has lived in Labrador since 1990, writing for NTV, CBC, the OKalaKatiget Society, and the Canadian Press. He has authored two other books and is a weekly columnist for Transcontinental Media.

references

St. John's *Telegram*

Melina & Keith II's captain fined $5,000, August 19, 2008

Makkovik searchers find body of snowmobiler, February 1, 2012

Lives trump savings: group, February 4, 2012

Conditions over Makkovik, helicopter maintenance impacted military response to search, admiral says, February 8, 2012

Conditions didn't prevent choppers from flying: company, February 8, 2012

Makkovik boy remembered at vigil, February 10, 2012

Unanimous support for SAR in house, March 8, 2012

Winters death leads to new SAR protocol, March 8, 2012

Winters SAR review report available, March 9, 2012

Teen's death leads to new SAR protocol, March 9, 2012

Ottawa announces third Griffon helicopter for 5 Wing Goose Bay, April 11, 2012

Questions linger over Burton Winters' death, May 8, 2012

Premier backs out of meeting with Winters family, May 10, 2012

Premier cancels meeting with dead boy's family, May 11, 2012

Premier won't hold "mini-inquiry" into death of Burton Winters, May 12, 2012

SAR protocols must change: Premier, May 25, 2012

Burton Winters' mother breaks her silence, May 29, 2012

Burton Winters story told through access to information: Wiseman, June 14, 2012

Battling for Burton, December 31, 2012

Body recovered from Labrador lake, May 9, 2013

CBC

Families sue feds, builders after Ryan's Commander deaths, October 18, 2006

Design led to Ryan's Commander disaster: report, November 20, 2006

Melina and Keith II sinking report urges reform, October 12, 2007

Runaway Labrador teen survived days on frozen island, October 6, 2011

Protesters try to send "search party" for MP Penashue, January 26, 2012

Search for missing Labrador boy finds tracks to open water, January 30, 2012

Missing Makkovik teen's snowmobile found, January 31, 2012

Missing Labrador teen snowmobiler found dead, February 1, 2012

Vigils held for Burton Winters, February 2, 2012

DND probes alleged delays in search for Labrador boy, February 2, 2012

Dead Labrador boy's family slams DND search "failure," February 2, 2012

Protesters demand search and rescue improvements, February 10, 2012

Federal minister joins call for Labrador search review, February 10, 2012

More Labrador vigils calling for better search and rescue, February 11, 2012

"He wanted to be home with us," family says of teen, February 13, 2012

Search for Labrador boy lost on ice raises more questions, March 22, 2012

Dunderdale demands answers from DND about Winters search, March 26, 2012

Calls continue for improved Labrador search and rescue, March 29, 2012

Penashue says Fifth Estate probe not "fair, full picture," March 30, 2012

Search for Winters was N.L. responsibility: MacKay, April 3, 2012

Liberals, Tories trade barbs over search inquiry calls, April 5, 2012

Labrador gets 3rd DND helicopter after Winters' death, April 11, 2012

Winters' family wants more Labrador search and rescue improvement, April 12, 2012

Winters protest moves to Penashue's office, April 19, 2012

DND recordings question N.L. role in search for Labrador boy, May 5, 2012

DND tapes question N.L. role in search for lost boy on ice, May 7, 2012

Burton Winters: RCMP, premier defend handling of Labrador teen's search, May 7, 2012

Searchers never recruited to look for Labrador teen on ice, May 8, 2012

RCMP criticized for role in search for lost Labrador boy, May 9, 2012

Burton Winters' family says Dunderdale nixed meeting, May 10, 2012

Premier says Winters meeting was turning into PR "stunt," May 11, 2012

Dunderdale slams feds for role in search for Labrador teen, May 22, 2012

Feds would co-operate with a provincial Winters inquiry, May 23, 2012

No inquiry on Winters search, despite federal co-operation, May 24, 2012

N.L. premier "at odds" with Peter MacKay, May 26, 2012

No problem with our role in Winters search says RCMP, May 28, 2012

Person rescued near cliff in Grand Falls-Windsor, July 5, 2012

Questions raised about Gander chopper availability, July 6, 2012

A year after tragedy, search-and-rescue doubts persist, January 29, 2013

Body of former Innu leader recovered from Labrador pond, May 9, 2013

Province mourns former Innu grand chief, May 10, 2013

Province reaches deal with air search and rescue volunteers, June 19, 2013

Canadian Press

Mounties searching for missing 14-year-old boy in northern Labrador, January 30, 1012

Body of missing boy found 19 kilometres from snowmobile, February 1, 2012

Death of Labrador boy raises questions about search and rescue, February 3, 2012

Weather slowed air search for Labrador boy, Forces say, February 4, 2012

RCMP denies report on lost Labrador boy, May 7, 2012

Cormorant should have flown in search: premier, May 23, 2012

Death of former First Nations grand chief prompts criticism of Canada's search-and-rescue efforts, May 13, 2013

Ottawa Citizen

Peter MacKay announces $8.1 million for search and rescue response, April 15, 2012

Peter MacKay relies on smoke and mirrors when it comes to an announcement of "new" search and rescue funding, say critics, April 15, 2012

NDP motion on search and rescue times defeated in Commons, June 14, 2012

Gander *Beacon*

In his memory, March 8, 2012

Maclean's

No search, no rescue, for Labrador's Burton Winters, February 20, 2012

The Aurora

Burton Winters remembered one year later, February 11, 2012

Call for independent investigation [letter to the editor], February 27, 2012

The Labradorian

Vigil held for Burton Winters, February 10, 2012

SAR protest prompts public meeting with Penashue, February 10, 2012

Uniting for Burton, February 20, 2012

Tribute to Labrador's son, February 24, 2012

"It's a start but not the end," March 19, 2012

Winters family and protesters sending a message: no more cover-ups, May 15, 2012

Province mourns former Innu grand chief, May 10, 2013

More concerns over search and rescue after Riche's death, May 21, 2013

Aboriginal Peoples Television Network

Family of dead Labrador teen wants meeting with Penashue, February 16, 2012

The boy who united Labrador, February 17, 2012

OKalaKatiget Society

Underwater camera to be used in search for missing boy, January 31, 2012

Labrador MHAs hosting vigil and rally for Burton Winters Search and Rescue Centre, February 23, 2012

National Post

Three military helicopters grounded in Labrador during search and rescue emergency, May 10, 2013

Globe and Mail

Nfld. Premier meets Harper in wake of search-and-rescue tragedy, September 24, 2012

The Nor'wester

Pilley's Island remembers Burton Winters, April 13, 2012